Benvenuto (Ed.)
Sustainable Green Chemistry
Green Chemical Processing

Green Chemical Processing

Edited by
Mark Anthony Benvenuto

Volume 1

Sustainable Green Chemistry

Edited by
Mark Anthony Benvenuto

DE GRUYTER

Editor
Prof. Dr. Mark Anthony Benvenuto
Department of Chemistry and Biochemistry
University of Detroit Mercy
4001 W. McNichols Road
Detroit, MI 48221-3038
USA

ISBN 978-3-11-044189-5
e-ISBN (PDF) 978-3-11-043585-6
e-ISBN (EPUB) 978-3-11-043350-0
Set-ISBN 978-3-11-043350-0
ISSN 2366-2115

Library of Congress Cataloging-in-Publication Data
A CIP catalog record for this book has been applied for at the Library of Congress.

Bibliographic information published by the Deutsche Nationalbibliothek
The Deutsche Nationalbibliothek lists this publication in the Deutsche Nationalbibliografie; detailed bibliographic data are available on the Internet at http://dnb.dnb.de.

Typesetting: Integra Software Services Pvt. Ltd.
Printing and binding: CPI books GmbH, Leck
Cover image: scyther5/iStock/Thinkstock
♾ Printed on acid-free paper
Printed in Germany

www.degruyter.com

Contents

List of Contributing Authors

Chapter 1
William F. Carroll, Jr.
Department of Chemistry
Indiana University
Bloomington, IN 47405
USA
wcarroll@indiana.edu

Chapter 2
Sarah A. Green
Department of Chemistry
Michigan Technological University
Houghton, MI 49931
USA
sgreen@mtu.edu

Chapter 3
Heinz Plaumann, Ph.D., FCIC
University of Detroit Mercy
4001 McNichols Av.
Detroit, MI 48221
USA
plaumahp@udmercy.edu

Chapter 4
Martin Straka
ÚJV Řež
Hlavní 130
250 68 Husinec- Řež
CZECH REPUBLIC
martin.straka@ujv.cz

Chapter 5
Lesley Putman*
Northern Michigan University
Marquette, MI 49855
USA
lputman@nmu.edu

Chris Nyland
Northern Michigan University
Marquette, MI 49855
USA

Kristine Parson
Northern Michigan University
Marquette, MI 49855
USA

Chapter 6
Larry Kolopajlo
Chemistry Department
Eastern Michigan University
Ypsilanti, MI 48197
USA
lkolopajl@emich.edu

Chapter 7
Matthew J. Mio
Department of Chemistry & Biochemistry
University of Detroit Mercy
4001 W. McNichols Rd.
Detroit, MI 48221-3038
USA
miomj@udmercy.edu

Chapter 8
Meghna Dilip
Department of Chemistry
Worcester State University
Worcester, MA 01602
USA

Margaret E. Kerr*
Department of Chemistry
Worcester State University
Worcester, MA 01602
USA
mkerr@worcester.edu

Mark Benvenuto

Where We Are and Where We Are Going with Green Chemistry

Abstract: A variety of green chemical principles and processes are discussed in this volume. They include how green chemistry has affected the petrochemical industry, as well as emerging, possible applications within it. Also included are several chapters that discuss how the 12 green chemistry principles are being applied in higher education (Twelve Principles of Green Chemistry, https://www.acs.org/content/acs/en/greenchemistry/what-is-green-chemistry/principles/12-principles-of-green-chemistry.html), when training students who will be the next generation of chemists and engineers, and how they are being examined in new ways in such processes as spent nuclear fuel remediation and the digestion of bisphenol A. In total, these chapters indicate that we have advanced significantly from the initial publication of what is now called the Brundtland report, and also imply that there are large areas in which improvement can still be made (*Our Common Future: The World Commission on Environment and Development*, 1987).

Introduction

Welcome to a new offering in the ongoing, widening discussion of green chemistry and green chemical processes. This volume is what we hope to be the first of many that will offer a forum to discuss where we and the science are, where such processes might be going, and where the need to improve is still evident.

One can hope that the now familiar 12 principles of green chemistry are well known and widely disseminated among not only chemists and chemical engineers, but among the wider, general public. When the discussion began over 30 years ago, there was no immediate consensus on what this move to improve what can now be called the traditional chemical enterprise should be titled or what its central tenets were to be, although thrusts toward sustainability and the use of the least amount of material were quickly established as aims and goals [1, 2]. Only later, after significant discussion, did the 12 principles emerge: (1) prevention of waste, (2) atom economy, (3) decreased toxicity in synthesis, (4) design of benign chemicals, (5) benign solvents, (6) energy efficiency, (7) use of renewable feedstocks, (8) reduction of derivatives, (9) increased use of catalysts, (10) design for biodegradation, (11) pollution prevention, and (12) accident prevention and benign chemistry [3]. Even today it would be presumptuous to think that the existing 12 principles are static, defined, and immutable. As new processes and techniques are developed, there may be reason to alter or add to what we now consider these basic tenets.

Since its inception, what has become the field of green chemistry may appear to have been constructed around and have a focus in the discipline of organic chemistry [2]. There seem to be several reasons for this. First, from the perspective of academic chemists, the professoriate, this may simply be because we teach it so much, not only to chemistry students but to biology students, engineering students, and to those students entering any of the health professions. Yet this point could be made about general chemistry classes, or at least the organic chemistry examples that are found in such. Second, from a broader perspective, this could simply be because we are now in an age that uses organics more than ever before, and in an amazingly wide array of applications, traditionally from petrochemical feedstocks or more recently from bio-based materials. The explosive growth of the plastics industry since roughly the end of the Second World War, an industry that is largely within the broader field of organic chemistry and that almost exclusively uses petrochemical-based starting materials, has brought with it an increased use of a wide variety of nonaqueous solvents, a large amount of waste material, and the need for more efficient syntheses. Third, it is quite possible that the early focus of green chemistry on organic chemistry is because the earliest champions for it were heavily involved in pharmaceuticals chemistry and organic synthesis, a field that has traditionally used a great deal of materials to produce what are referred to as small volume, high value end-products – namely, drugs [4–7]. Fourth, it may very well be that since the chemical industry utilizes a large number of organic chemists at each degree level (bachelor's, master's, and doctorate), this is what most large academic institutions' chemistry departments produce in abundance. Since this has gone on for several decades, one can argue that organic chemistry has seen heavier, more detailed study in the course of that amount of time than some other disciplines.

This volume does discuss some organic and biochemical topics, but discusses some rather nonorganic topics as well, the latter of which can be viewed as a very broad area still relatively untouched by those wishing to improve chemical processes, the chemical industry, and chemical education. For example, while the production and use of plastics has been seriously examined for decades in terms of greening at the manufacturing and production stage as well as at the recycling stage, metals and their production are often considered "ungreenable" by many because carbon dioxide is coproduced stoichiometrically when carbon is used as a reducing agent – and always has been. Thus, what might be called efforts at greening metals chemistry have been focused at the end (or we might say, confined to the end), at re-use and recycling. But this is simply what has worked in the past, and what has been economically the most feasible and easiest means of increasing efficiency in the life cycle of a metal material, be it element or alloy. Past chemical practices do not dictate what must be done for the future. Even though efforts at greening these old, established processes may seem impossible at the moment, revolutionary, transformative chemistry and processes may emerge in the future if there is some incentive for such improvements.

This Volume

Green Chemistry in the Chemical Industry

We are honored to have Dr. Bill Carroll, former American Chemical Society President as well as former ACS Chairman of the Board, who wrote a chapter on green chemistry and chemical industry. Bill's perspective may actually be unique, as it appears that no one else has held both these positions and been a leader in the chemical industry for well over three decades. Dr. Carroll examines the principles from an economic perspective, discusses how the cost of energy is factored in, looks at the historical underpinnings of industrial chemical processes, and illustrates how all this is intertwined with the underlying chemistry. We can all learn from his insights and what he has to offer.

Progress and Barriers

It is Professor Sarah Green who provides this volume with what may be one of the most perceptively written, broad assessments of green chemistry and where it fits in amidst our chosen discipline. Her chapter is obviously the result of a long and very intelligent, well-reasoned assessment of the principles of green chemistry and how they drive us forward as chemists and as people. She also points out challenges that we can hope chemists and chemical engineers will meet in the future.

Switchable Polarity Solvents

Dr. Heinz Plaumann brings almost three decades of experience in industry – much of it at BASF – to the very interesting chapter he has authored on what are called switchable polarity solvents, meaning materials that can undergo a change in polarity depending on conditions. These materials are relatively new, and thus have significant potential in a variety of chemical reactions. Solvents that can undergo a reversal of polarity under different conditions, as these do, may also serve as a significant improvement for one or more significant reactions, reaction types, or types of material recovery, and therefore become a step toward greener syntheses.

Actinide Recovery

A fascinating chapter on actinide recovery and the use of ionic liquids for that end has been written by Dr. Martin Straka. The potential for the use of ionic liquids when it comes to the recovery of these rare and potentially useful elements is certainly great, as significant effort and energy are put into the original refining of elements such as uranium and thorium, the former for use as a nuclear fuel. This paper examines how actinides interact with ionic liquids in terms of solvent stability and chemical reactivity.

Phytodegradation of Bisphenol A

An excellent example of a process that is original and that can be considered out of the main stream – a form of phytoremediation – is presented by Dr. Putman and her colleagues, in their examination of a novel means to degrade the ubiquitous bisphenol A (BPA) molecule from effluent streams. Putman indicates in her introduction that the quality of life that people enjoy today is afforded in part by modern plastics and is not something that people will willingly surrender. Thus, the chapter discusses a novel means of removing BPA from effluent streams utilizing little bluestem seeds. The technique definitely is a green approach to a large-scale problem and is new enough that we hope a wider adoption of it will occur in the future.

Green Chemistry in K-12 and Higher Education

Professor Larry Kolopajlo presents a very thorough examination of how green chemistry and its principles have been applied in higher education, especially in chemical education. His chapter examines in detail how and where green chemistry is taught in high school as well as at the college level.

Green Chemistry for Chemistry Majors Versus Nonscience College Students

Professors Kerr and Dilip approach the topic of green chemistry in the curriculum in a way that may be substantially different from many others. They examine the topic in terms of our students, but also in terms of the much broader audience of college students inclusive of those who are not pursuing chemistry as an academic discipline. As well, they discuss how the subject matter can be taught using an online delivery.

Greening the Organic Labs

Professor Matthew Mio has contributed a chapter that delves into the greening of organic chemistry labs that are taught at the undergraduate college level, both in terms of teaching the students the 12 principles and in terms of the economic incentive at a university. His paper does an excellent job of illustrating how such improvements are made, and how they are integrated into the students' learning.

Summary

The chapters included in this first volume represent a wide variety of ways in which green chemical principles are being applied, whether such applications are in industry, in research, or in academia. We recognize that some of the subjects presented here are rather distant from what have become the current main thrusts of the field, such as the well-established different types of improvements in organic syntheses.

But it is believed that the 12 principles of green chemistry can and will find uses in even broader perspective than those presented here. It is hoped that this volume can become an outlet for such ideas and research, that it makes a valuable contribution to this growing and important discussion, and that the future will see the implementation of what has been presented here.

References

[1] Our Common Future: The World Commission on Environment and Development. Oxford University Press, 1987, 019282080X.

[2] Trost, B. M. The Atom Economy – A Search for Synthetic Efficiency. *Science* 1991, *254*, 1471.

[3] Twelve Principles of Green Chemistry. https://www.acs.org/content/acs/en/greenchemistry/what-is-green-chemistry/principles/12-principles-of-green-chemistry.html (accessed June 17, 2016).

[4] Anastas, P. T.; Warner, J. C. Green Chemistry: Theory and Practice. Oxford University Press, 2000, 9780198506980.

[5] Anastas, P. T.; Kirchhoff, M. M. Origins, Current Status, and Future Challenges of Green Chemistry. *Acc Chem Res* 2002, *35*, 686–94.

[6] Anastas, P. T.; Bickart, P. H.; Kirchhoff, M. M. Designing Safer Polymers. Wiley-Interscience: New York, 2000.

[7] Tundo, P.; Anastas, P.; Black, D. S.; Breen, J.; Collins, T.; Memoli, S.; Miyamoto, J.; Polyakoff, M.; Tumas, W. Synthetic Pathways and Processes in Green Chemistry: Introductory Overview. *Pure Appl Chem* 2000, *72*, 1207–30.

William F. Carroll, Jr.

1 Upon Further Review: A Commodity Chemist on Green Chemistry

Abstract: Green chemistry is most often thought of in the context of specialty or pharmaceutical chemicals where many synthetic chemistry approaches are in play. However, principles similar to those of green chemistry and engineering were employed over the years in reducing cost and increasing volume of chemicals that became commodities. This paper considers some of those principles, their impact, and some perspectives on the potential and limits associated with green chemistry for commodity chemicals.

Keywords: Green Chemistry; Commodity Chemicals; Life Cycle Analysis

During the year I was president of the American Chemical Society (ACS), 2005, I was the head of our delegation visiting China. It was a joy to return to see the progress the Chinese people had made since my first trip on behalf of my employer, nearly 20 years before. In addition to the challenge of a week of eating two banquets a day, I had the privilege of meeting literally hundreds of bright and serious students and professors. It was a wonderful time.

One particular exchange stood out. At Tsinghua University in Beijing, we had an interesting and wide-ranging roundtable discussion with students and professors about chemistry and industry. Now, while the situation there may be different in 2015, during that discussion in 2005 one student asserted, "Of course we know that pollution is inevitable with growth and progress."

I was stopped. In a moment, the history of the past 30 years of the chemical industry as I have known it went through my mind. I thought about the advent of voluntary chemical and safety programs like Responsible Care® and OSHA Star, and the difference they made in my company and in the industry. I thought about the progress we have made in our own universities in educating students in the principles of green chemistry. To tell the truth, when I was in that student's place 30 years ago, I might have made the same statement. But 30 years later it seemed unimaginable.

You see, we have changed innumerable things in the practice of chemistry, but the most important thing we have changed is our minds. I couldn't let that student's statement pass. I said,

> "Please forgive me, but I need to disagree. From the perspective of the chemical industry, pollution and progress are not synonymous. Pollution is waste and waste is cost. Progress is benefit and a better quality of life. The two things are inherently different. That understanding is fundamental to what we call Green Chemistry. Your job as a chemist is to solve problems, to be sure. However, it is not to find a solution to a technical problem, but to challenge yourself constantly to find a better solution – that is, more benefit and less cost. I suggest you visit the ACS website and learn about the Green Chemistry Institute (GCI)."

This student, before even entering the working world of chemistry, had resigned himself to defeat. It shows the depth of the misconception, the difficulty of changing minds, and the importance of the principles of green chemistry in remediating this perception problem.

To me, the message of green chemistry and engineering is whether your particular point of emphasis is environmental, as in minimizing pollution, or economic, as in minimizing cost, the principles are important tools for accomplishing your goal. And frankly, I've found that once made aware of them, my fellow chemists and engineers working in industry find most of them nearly intuitive.

What I've just described to you are the technological underpinnings of green chemistry as I see them. We in ACS have always viewed it as a process of innovation, driven by scientists in discovery mode. We have been clear: green chemistry is not a regulatory program, because regulation tends to narrow the field of thought and not expand it. On a previous incarnation of the front page of the Green Chemistry Institute (GCI) website, it said:

> Green Chemistry differs from previous approaches to many environmental issues. Rather than using regulatory restrictions, it unleashes the creativity and innovation of our scientists and engineers in designing and discovering the next generation of chemicals and materials so that they provide increased performance and increased value while meeting all goals to protect and enhance human health and the environment [1].

As I write this, it is 10 years later. I am in my last week of a 36-year career with a commodity chemical manufacturer. Over that time I've learned that "The Chemical Industry" is not one thing: commodities are different from specialties, which are different from fine chemicals or pharmaceuticals. Each segment views its business differently because the costs and prices and products and customers are different.

I'd like to expand a bit about two of these segments, commodities, and specialties. Commodity chemicals are manufactured in large volumes. Typically, most of the cost associated with a commodity is raw material cost or the capital cost of the plant. Relatively less of the cost of a commodity is tied up in labor, advertising, or R&D.

The main thing that distinguishes a commodity from a specialty is that commodities are fungible; that means that one company's version is indistinguishable from another company's version. Commodities are difficult if not impossible to brand or differentiate. These things are directly the opposite of specialties, which can be custom-made, distinguished by manufacturer, branded, and sold at a premium price because of the differentiation.

So why would someone whose career was spent in the commodity chemicals industry be writing a paper about green chemistry? Perhaps I have a different and counterintuitive perspective on the benefits of green chemistry. In this paper I'd like to make three points.

First, many commodities – certain chemicals produced at low relative cost and high volumes – got that way because of chemistry and engineering operating principles that are indistinguishable from at least some of the principles of green chemistry and engineering. These principles were used to take cost out of chemical manufacturing.

The second is similar. Done correctly, the benefit of green chemistry, particularly in commodity markets, may be as much economic as it is environmental, although both are likely to occur simultaneously.

Finally, green chemistry's greatest potential impact is on energy.

Anyone who has spent time with the principles of green chemistry and green engineering knows it is difficult if not impossible to optimize all of them simultaneously in the preparation of a chemical or product. In manipulating conditions, there is a need to simultaneously consider a number of response variables – from energy use, to greenhouse gas generation, to chemical hazard or even cost – even as the principles provide guidance on how to proceed. The principles do not always point in the same direction for all responses, which become a dilemma for those who are engaged in what is known as alternatives assessment: devising means to eliminate chemicals of concern while minimizing process impact and providing efficacy at a cost the market can afford.

I've always viewed the principles of green chemistry and green engineering as checkpoints in chemical and process design. For me, the principles read: "All things being equal, consider this before considering that." Not absolute rules, but points to be taken on board. And when they send you different directions, choices must be made.

Here are some thoughts on the relationship of commodity chemicals to green chemistry.

1.1 Commodity Chemicals Got That Way Through the Use of Principles Identical to Many of Those of Green Chemistry and Engineering

Perhaps the most commodity of all the commodity chemicals is ethylene. Hundreds of billions of pounds of it are made globally every year. It is the basis of most of the commodity polymers. It is also a natural product that enhances fruit ripening. And for all the volume we produce today, when the polymer industry was new, ethylene was a challenging material to make.

In the 1920s, ethylene was made by dehydrating ethanol, which was obtained from fermentation. Prior to the development of the polymer industry, most of that ethylene was reacted with bromine to make ethylene dibromide, an additive in leaded gasoline. For other industrial processes, when a two-carbon unit was needed, it could be obtained by functionalizing acetylene, which was made by adding water to calcium carbide.

As automobiles became more popular, engineers at Dow Chemical Company started working on a process to make ethylene from naphtha, a petroleum distillate [2]. Availability of ethylene increased with the more robust process and a more abundant raw material for its synthesis. Today, in many countries, naphtha is the raw material of choice for ethylene. Most of the ethylene in the United States is made by dehydrogenating ethane, harvested from natural gas.

Consider some green chemistry implications of these three processes. Calcium carbide is made from coke and limestone, heated in an electric arc furnace at 2,000°. The ovens are highly energy intensive, and the process itself (as well as coke production leading to it) can be extremely dirty. After making acetylene, two to three times the mass of acetylene is left behind as calcium hydroxide waste. Ultimately, this material can be dehydrated to make lime and used in making cement if the volume and location matches, but it is yet another process.

Dehydrating ethanol is somewhat less daunting, but was limited by the capacity to produce it, particularly in the 1930s. Then, dehydration was catalyzed by sulfuric acid. Today, it can be accomplished thermally at ca 250 °C over a zeolite catalyst. By the nature of a dehydration reaction, 100 % molecular conversion is only 61 % mass conversion.

The steam-cracking process utilizes a hydrocarbon raw material, generally ethane or other liquefiable hydrocarbons separated from natural gas. Steam is both a heating source and a diluent; the process takes place at about 850 °C, quickly, and with a fast quench, maintaining appropriate dilution and kinetic control. Steam cracking of ethane yields ethylene and hydrogen from nearly 80 % of the starting material. Minor side products include methane, propylene, butadiene, and other C4s, all useful hydrocarbons. There is virtually no waste [3].

The steam-cracking process is the dominant process in the world. It became so because of the difficulty, resource use and high cost associated with the other processes. For local reasons, calcium carbide for acetylene production is done industrially only in China; producing ethylene from ethanol is done for marketing purposes to make plastics from biologically based materials.[1]

From the first indication that steam cracking could produce ethylene, engineering principles congruent with many of the principles of green chemistry came into play, and at maturity this process scores fairly well on a lot of them. Nearly perfect

1 In 2011, Dow and Mitsui announced a project to produce 320,000 MT ethylene and ultimately polyethylene, from 720,000 m^3 ethanol. This plant would require over 100,000 ha (250,000 acres, ca 400 sq mi) sugarcane production. (Dow-Mitsui Investor Relations Presentation). In 2013, the polyethylene plant was put on hold, with ethanol production still planned. There are few ethanol-to-ethylene facilities, although one owned by Braskem in Brazil is said to be cost competitive with a (relatively expensive) naphtha-based plant. Bio-based polyethylene is said to command a "premium of 40 % or more from clients eager for an enviro-marketing edge." Plastics News, January 10, 2013, http://www.plasticsnews.com/article/20130110/NEWS/301109988/dow-and-mitsui-postpone-sugarcane-polymer-plant. Accessed September 27, 2015

score on preventing waste, atom economy, and avoidance of solvents and derivatization, particularly compared to the carbide process. Neither ethane nor ethylene itself is highly toxic, and the steam-cracking process uses little if any toxic ingredients.

On the other side of the coin, some of the principles are challenging. Since the target is a two-carbon hydrocarbon, there's not much that can be done to reduce the inherent flammability or explosivity of the material. The process is done at high temperature and pressure and carries an energy cost; it is difficult to imagine the reaction occurring at ambient conditions.

Designing for degradation is a function of the downstream products, so all three processes fare the same. While the carbide process is clearly inferior on environmental (and cost) grounds, life cycle assessment is appropriate to score the other two, comparing ethanol production and taking into account atom economy of the two processes. If, however, production cost could be used as a crude surrogate for resource use, the ethane steam-cracking process is less costly.

Perhaps the principle of green chemistry that is most daunting for commodities is the preference for renewable resources, which in this case I take to mean bio-based raw materials. I see three challenges in this regard, which I will abbreviate as scale, substitution, and life cycle considerations.

1.1.1 Scale

In 2014, about 14 billion gallons (92 billion pounds) of ethanol was produced in the United States, mostly from corn. At 100 % yield that would produce about 60 billion pounds of ethylene, similar to the US production. If all US ethylene were to be made from ethanol, even though the raw material supply is evenly matched, the dehydration plants to accomplish this would have to be built, and because of the physical properties of ethylene they would need to be built with pipeline access to downstream users, and ethanol would have to be transported to them.

But the biggest question: all of that agricultural production would go to manufacture just one commodity chemical. The United States manufactures about 25 % of the world's ethylene, and whether there could be bio-based raw material in the other countries of the world is unknown.

1.1.2 Substitution

The National Academies report "Sustainability for the Nation: Resource Connection and Governance Linkages" [4] discussed the magnitude of the US government program, subsidizing ethanol production and the impact it had on the marketplace. The policy subsidizing fuel use of ethanol had the unintended consequence of pushing Midwestern farmers out of soybeans and into corn. Soybeans became short and induced farmers – largely in Brazil – to plant soybeans to replace lost volume.

Connections that are indirect can nonetheless be highly significant. Demand for ethanol in the United States caused the price of corn to rise and caused a shift in land use from soybean production to corn production. To fill the void, land was deforested in other countries and planted in soybeans.

Accepting that corn is a poor agricultural feedstock for ethanol, similar substitution consideration would have to be carried out regarding the advisability of ethanol-to-ethylene even if derived from higher sugar-yielding crops. The Dow-Mitsui plant was designed for 720,000 m^3 ethanol /114,000 ha sugarcane (ca 2 tonne/acre). The US ethylene capacity is about 25 million tonnes and would at best require significant redeployment of land – on the order of thousands of square miles – to sugarcane, or a switch of ethanol use from fuel to chemicals, presumably replaced by fossil products.

Early in this century, a similar dynamic drove the palm oil market – approximately the same size globally as US production of ethanol – into unsustainable land use, particularly with respect to indigenous animals and the risk of an agricultural monoculture. The situation has improved with acceptance of principles developed by the Roundtable on Sustainable Palm Oil [5]. It is difficult to know what impact another program of similar size would have on agriculture, particularly if farmland currently producing food were taken out of service in favor of a crop designed to produce raw materials for commodity chemicals globally.

1.1.3 Life Cycle Considerations

In order for a chemical raw material to be truly sustainable, it should be a material that is sustainably grown specifically for this purpose or is derived from a material not currently being used productively. Producing ethylene from corn stover cellulose, which appears to be a waste, seems like a good idea, but those corn stovers are currently plowed under for soil amendment. Starting from a crop that is purpose grown – switchgrass or another "fallow field" product – or repurposing of truly waste cellulose headed for landfill (and not paper recycling) might pass the life cycle test at some scale. Whether it could do so at tens of billions of pounds is an open question.

Two of the winning entries of the 2015 Green Chemistry Challenge awards involved the use of industrial off-gases or CO_2 as feedstock. The first innovation routed off-gases from steel mills to genetically modified organisms capable of converting them to ethanol, acetic acid or 2,3-butanediol. The second involved a new photobioreactor for fixing CO_2 via blue-green algae. Either would fit the requirement of turning not just a waste but a pollutant into necessary products.

However, 37 years of experience asks whether such processes will scale to the kinds of volumes we're talking about, and if so, will they be cost competitive, both fixed and variable? It would be wonderful if the answer were yes on both counts. But for now, bio-based solutions for the chemical industry will work best where the amounts are manageable, genetic modification in a factory environment is tolerated (1,3-propanediol converted by bacteria from glycerol, for example) or where the chemistry takes advantage of a unique molecular scaffold produced biologically.

1.2 The Benefit of Green Chemistry May Be as Much Economic as It Is Environmental, Although Both Are Important

In the book *Eco-Efficiency*, Frank Popoff and Livio DeSimone, two former chemical industry CEOs, described the perhaps counterintuitive notion to some that when you reduce pollution you often times also reduce costs – although sometimes they are hidden costs that are not always fully recognized by our accounting system [6]. And, conversely, a responsible process of cost reduction generally results in greater efficiency and almost inevitably means a reduction in waste or pollution at some point in the life cycle.

The key is in optimizing those two variables simultaneously. It is, of course, possible to design more environmentally sound processes that are so costly that they will never be commercialized. Similarly, it is possible to design processes that are "lower cost" but so ignore safety and environmental practices that they are actually not "economical." Just as I advised the Chinese student, more creativity is needed so as to not just find something that works, but to find better solutions on all counts.

John Warner puts it this way: "Green Chemistry really provides a 'holy grail' of sorts: better performance, better cost and 'Oh by the way … it is better for human health and the environment.' Given these criteria, what market barrier exists for adopting a greener product save its invention in the first place?" [7].

The GCI, which is a part of ACS, operates a number of industry roundtables, where companies in a particular market space can gather and discuss greener chemistry in a pre-competitive environment. The most successful of these so far is the Pharmaceutical Roundtable [8]. Much of the discussion in the roundtables includes process characteristics: as an example, distillation as a means of separation is ubiquitous. Could less energy-intensive processes be developed? In addition, the roundtables seek simpler synthetic methods that utilize less toxic materials, less organic solvent or fewer waste-generating steps. In either case, while there may well be environmental benefit from reduction of waste generation and disposal, reduced raw material and waste disposal costs are at least a bonus. Other roundtables or other companies that find a similar economic benefit to the use of green chemistry will also be successful, particularly those in cost-sensitive industries.

1.3 Green Chemistry Can Have Its Biggest Impact if It Enables Green Energy

I've just described an appeal of green chemistry: its perceived potential to reduce resource use and waste. But it is important to put the magnitude of such potential in perspective. The use of coal, oil, and natural gas for fuel is a factor of 10–20 higher than their use for both raw materials and energy in the chemical industry. Even if green chemistry were able to increase the efficiency of the entire chemical industry by 10 % – a huge goal – the overall magnitude of its impact would be less than 1 % on total resource use.

On the other hand, studies of energy efficiency show that only about 45 % of the fuel burned for energy actually gets put to use [9]. The remainder – particularly from the thermal processes that generate electricity – goes to waste heat. One percent decrease in total resource use via green chemistry would be important, but not a game changer.

Some will take this as a cynical statement by a grizzled old denier. Nothing could be further from the truth. The point is: Green chemistry can make wonderful contributions to a company's bottom line by reducing waste, particularly hazardous waste, and by simplifying synthesis, which translates to eco-efficiency. Green chemistry thinking can mitigate potential accidents locally by designing safe processes, choosing materials with a better hazard profile, and mitigating environmental harm, particularly from irresponsible waste treatment. But if green chemistry is truly to make a difference to society writ large it will be because of the contribution it makes to energy efficiency and ultimately use.

Consider the barriers to greater adoption of renewable energy, particularly wind and photovoltaic solar. In both cases, potential uptime has to be taken into account in ways that do not need to be considered for conventional energy. The wind doesn't blow all the time, nor does the sun shine; thus, wind and solar facilities need to be sized at a capacity that is a multiple of conventional plants, and conventional energy generation must be in place to "back up" renewable facilities. On the other hand, contributions made by chemists in the field of energy storage and collection efficiency can allow resizing of such facilities to generate more during peak productive hours with reliable storage and retrieval during nonproductive hours.

Chemistry also could also contribute in harvesting waste heat. In these days, when chemical companies build electrical capacity for their plants they almost always build a combined heat and power facility that is capable of using low-pressure steam or hot water for local heating needs. These plants reach 60–70 % efficiency rather than 30–40 %. Thermoelectric materials if developed could be retrofitted into existing plants to harvest heat from low-pressure steam or hot water. And the real holy grail for chemistry is storing nontraditional energy by splitting water into hydrogen and oxygen, and mimicking natural photosynthesis by using the hydrogen as a reducing agent for carbon dioxide to boot.

1.4 For the Future

The big victory is somewhere still in the future, but the principles of green chemistry can serve the industry well in some way at all scales. The most elegant and interesting may be in the specialty, fine or pharmaceutical area, but solid, if unobvious contributions have been and will be made in commodities as well.

References

[1] Green Chemistry Institute. http://www.acs.org/content/acs/en/greenchemistry.html (accessed September 23, 2015).
[2] Bohning, J. J. Styrene at Dow for the World War II Synthetic Rubber Program. *Chem Heritage* 2004, *22* (3), 8.
[3] Chauvel, A.; Lefebvre, G. *Petrochemical Processes: Technical and Economic Characteristics*. Gulf Publishing Company, Houston, Tex.; French edition: Paris: Éditions Technip, 1989.
[4] National Academies Press. Sustainability for the Nation: Resource Connection and Governance Linkages, http://www.nap.edu/read/13471/chapter/3#21 (accessed September 23, 2015).
[5] Roundtable on Sustainable Palm Oil. http://www.rspo.org/ (accessed September 23, 2015).
[6] Popoff, F.; DeSimone, L. *Eco-Efficiency: The Business Link to Sustainable Development*, MIT Press, Cambridge, MA, 2000, ISBN: 0-262-04162-6.
[7] Warner, J. C. Where We Should Focus Green Chemistry Efforts. *Aldrichimica Acta*, 2015, *48* (1), 29.
[8] American Chemical Society. Green Chemistry Institute, Pharmaceutical Roundtable, http://www.acs.org/content/acs/en/greenchemistry/industry-business/pharmaceutical.html (accessed September 23, 2015).
[9] Lawrence Livermore National Laboratory Energy Flow Diagram. https://flowcharts.llnl.gov/energy.html (accessed September 27, 2015).

Sarah A. Green

2 Green Chemistry: Progress and Barriers

Abstract: Green chemistry can advance both the health of the environment and the primary objectives of the chemical enterprise: to understand the behavior of chemical substances and to use that knowledge to make useful substances. We expect chemical research and manufacturing to be done in a manner that preserves the health and safety of workers; green chemistry extends that expectation to encompass the health and safety of the planet. While green chemistry may currently be treated as an independent branch of research, it should, like safety, eventually become integral to all chemistry activities. While enormous progress has been made in shifting from "brown" to green chemistry, much more effort is needed to effect a sustainable economy. Implementation of new, greener paradigms in chemistry is slow because of lack of knowledge, ends-justify-the-means thinking, systems inertia, and lack of financial or policy incentives.

Keywords: environment, green chemistry, policy, safety, sustainability

When I first encountered the term green chemistry as a graduate student, I naively decided that it was too late for me to commit my career to the topic because the concept was so obvious that by the time I graduated it would be thoroughly embedded into the enterprise of chemistry. Sadly, the world moves at its own pace and the common sense of green chemistry that seemed so obvious to me in the 1990s has taken decades to penetrate into the daily life of chemists. At least now the idea of green chemistry is broadly recognized, although still very far from universally implemented. Reaching this level of penetration has required hard work by many people and organizations. Looking back from a few decades of experience, I can better identify some of the reasons that change is slow, as well as see the successes since the term green chemistry was introduced.

To understand the trajectory of green chemistry we can consider how it fits into the traditional practice of chemistry, how it aligns with our ideas of the progress of chemistry, and what factors impede its more rapid implementation.

Green chemistry, as defined by the 12 principles [1], has incorporated techniques from many branches of chemistry, as well as from other areas of science and engineering, and has spawned entirely new lines of inquiry. There can be no doubt that the increasing obligation to consider environmental impacts in all human endeavors, including chemistry, has spurred advances in catalysis, toxicology, processes chemistry, safety procedures, separations, biotechnology, novel feedstocks, analytical methods, synthesis, and many others.

A recurring question in the research domain is whether green chemistry constitutes its own branch of chemistry like the traditional divisions of organic, inorganic, analytical, and physical chemistry, or is a guideline defining how *all* chemistry should be performed, akin to safety and ethical standards. I suggest that currently it is the former, but ultimately it needs to become the latter. At its current stage of development, both technical and cultural advances are required before green chemistry can become thoroughly embedded into all chemistry endeavors, as it ultimately should be.

Currently, green chemistry is not yet mature enough to be automatically integrated as a baseline expectation into every activity. Research focused specifically on the topic remains essential because the chemistry community is still inexperienced in the holistic thinking that is required to assess processes according to green chemistry principles. Additionally, many tools and techniques that will increase the "greenness" of existing processes are yet to be developed. More excitingly, entirely new inventions and creative strategies are emerging that will supersede current practices to make this transition a reality.

Eventually, green chemistry should evolve from an independent branch of science into a guideline, like safety, that is incorporated into everything a chemist does. While we still need experts in chemical safety we expect that all chemists know and implement safe practices in all their work. The parallels between laboratory safety and green chemistry are clear. Consider, for example, the basic safety goal of preventing chemical explosions in research laboratories. Fundamentally, achieving that goal requires a thorough understanding of the underlying chemistry of explosions (thermodynamics, gas evolution, ignition, vapor pressures, etc.), the ability to anticipate the formation of explosive mixtures, and the technical means to handle and manage potentially explosive substances. A more advanced effort assesses whether the explosive substance is really essential to the final goal and redesign products or processes to eliminate the hazard altogether. Equally important for safety are changes in laboratory culture that enable practitioners to recognize that explosions are not an inevitable result of chemical procedures, learn to anticipate dangerous situations, and move to engrain safety consciousness into every stage of their activities. A similar progression can be seen in greening a chemical process. The preliminary step is to understand the underlying chemistry of the environmental risks of substances employed (fate and transport in the environment, degradation pathways and rates, toxicology, greenhouse gas potentials, etc.). Then one considers how to manage, minimize, and mitigate those environmental risks. Finally, processes and products may be entirely redesigned to achieve the desired goals while eliminating hazards.

Surprisingly, the parallels between safety and green chemistry are rarely articulated. Green chemistry is really the next step in safe practices, extending the net of safety beyond immediate laboratory occupants to encompass global populations and ecosystems. Yet often teams in a chemical enterprise focused on safety have little interaction with those concerned with green or sustainable chemistry, and they typically don't recognize their overlapping goals.

2.1 Brown versus Green Chemistry: Aligning the Goals of Traditional and Green Chemistry

To develop a sustainable chemistry framework, it's helpful to consider what chemists do and how they define progress. To achieve a sustainable chemistry enterprise we need to recognize that advances in green chemistry simultaneously advance chemistry as a whole.

Chemistry can be broadly divided into two primary activities: understanding things and making things, both contingent on a third, measuring things. All have their roots in the "brown" chemistry of the industrial revolution [2]. Yet a sustainable future requires that all three of these endeavors transform from brown to green. This stance is not anti-chemistry; instead it is a challenge to do new and better chemistry. In fact, greening is synergistic with the fundamental pursuits of chemists: understanding, making, and measuring. More sustainable chemistry requires a deeper chemical understanding, better ways of making, and improved measurements. The fundamental philosophy of chemistry is solving problems with molecular tools; green chemistry insists that a molecular approach to problem solving does not entitle us to be blind to the global impacts of the solutions.

As a primary goal chemists aim to understand the molecular basis of the world, including biotic and abiotic phenomena in nature, human-made materials, and the interactions among them. Our drive for understanding extends from molecular to macroscopic scales and asks how the first defines the second. Chemists want to know how atoms are transferred between molecules and into materials, and how energy is transferred through molecular and atomic processes. We want to understand how pharmaceutical molecules interact with human biology, how soil microbes manipulate nitrogen compounds, how paint reacts to photons, how ions travel through membranes, how electrons align with fields, and how individual atoms contribute to nanoscale properties, and infinite other fundamental questions. The search for chemical understanding isn't inherently brown chemistry, although many of the questions chemists have sought to answer in the past 100 years were driven by the technologies of the time: they knew how to apply heat and pressure to macroscopic systems and measure the results.

The general public is perhaps most familiar with the second aim of chemists: making things. We want to find new ways to make substances that nature has invented and make new substances that are unknown to nature. Both are about reorganizing atoms and molecules in a deliberate fashion. Traditional chemistry relied on a few brute force methods described in first year chemistry textbooks: heat to break bonds, pressure and heat to shift equilibria, and manipulating concentrations and solvents to modify reaction rates. Only with the (usually brief) introduction to catalysts do first year students glimpse the fact that the classical two-dimensional reaction coordinate diagram is not etched in stone and that clever chemists can circumvent the activation energy mountain. Until very recently, chemists have relied largely on brown chemistry to make nearly every one of the over 100,000 chemical substances in use today [3].

We enhance our abilities to both understand and make substances through the development of new analytical tools and techniques that allow us to map the molecular and atomic world at ever-finer detail and to make, purify, and identify substances never seen before. Analytical chemistry has not traditionally focused on green processes. Indeed many of the U.S. Environmental Protection Agency (EPA) methods for measuring pollutants use large volumes of toxic organic solvents, and often include organic or metal reagents that may also be environmentally unfriendly [4]. Of course, the three main chemistry goals of measuring, understanding, and making are not independent. Rather they are iterative. When we understand we can better make and vice versa. Green chemistry has the potential to advance all of these activities by framing them within the global context of improving human and planetary health. By analyzing each effort at multiple scales, green chemistry aims to avoid implementing a solution to a narrow problem that in fact creates more or bigger problems elsewhere.

As a starting point, we can examine the intersection between progress in chemistry as traditionally viewed and the goals of green chemistry. In our search for chemical understanding, progress can be defined as an enhanced ability to predict, design, control, and anticipate chemical behavior. This fundamental understanding is an essential component to the practice and theory of green chemistry. Indeed many chemists who might initially deny that they practice green chemistry actually contribute substantially to the field. For example, computational chemistry is perhaps the ultimate waste prevention strategy. Our increasing ability to understand and predict chemical behavior in silico reduces waste of both materials and time because uncountable unproductive experiments are avoided while chemists focus on the most promising pathways to successful synthesis. Similarly, nearly every chemist who studies catalysis contributes to the field of green chemistry because the goals of developing catalytic materials with enhanced selectivity and durability, improved efficiency, mild reaction conditions, and reduced toxicity fit the very definition of green chemistry [5, 6]. Understanding toxicity at the molecular level is essential for both green chemistry and to advance the ancient goal of employing chemistry to better human health and well-being. Both green chemistry and health-related chemistry demand an understanding of how exogenous molecules interact with human biology. Thus, there is strong synergy between green chemistry and the long recognized aspiration of chemists to understand how the world works. An effort to make that link explicit will both expand our chemical understanding and advance environmental health.

While some chemists are doing green chemistry without realizing it, others may think they are doing green chemistry without the necessary global perspective. For example, chemists who are specifically pursuing chemistry for sustainability may assume they are engaged in green chemistry. However, green chemistry requires more than a green goal such as better batteries, fuel efficiency, or material recycling. It requires that the means to the end be conducted with minimal environmental impact and that the entire lifecycle of the substances be considered. A notorious example is the widespread addition of methyl *tert*-butyl ether to gasoline to reduce tailpipe emissions without considering its potential to contaminate groundwater. To avoid this type of "solution" it is essential to comprehensively consider impacts at all scales from the laboratory to the planet.

Understanding chemistry on a planetary scale is essential to avoid dangerous global experiments such as the infusion of ozone-destroying chlorofluorocarbons into the stratosphere. We now have the ability to anticipate ozone destruction or global warming impacts of gaseous compounds released to the atmosphere and promote policies to limit their releases before planet-wide effects are felt. Assessment of ozone-depleting substances is ongoing by the Montreal Protocol, which now evaluates global warming potentials of halogenated compounds [7]. Although enormous advances are being made, a similar depth of understanding and predictive ability is lacking for less volatile synthetic compounds that are widely dispersed in water and soil, much more complex environments. An enormous challenge is tracking and assessing the impacts of molecules that move among air, water, and solid phases, are taken up by biota, and undergo transformations to new substances along these pathways. Anthropogenic molecules are now intermingling with the vast reservoir of natural materials on the Earth with unknown consequences for the biosphere. Much progress in understanding fundamental chemistry has been driven by our need to understand chemical processes in the environment. An ongoing dialogue between laboratory chemists and those engaged in understanding Earth's chemistry is required to truly fulfill the mandate of green chemistry.

In the more concrete chemical enterprise of making things, the first measure of progress is typically defined as increasing efficiency, measured in material, energy, time, and cost. In that realm progress is fundamentally aligned with green chemistry, and indeed many successes have been driven by efforts to reduce waste disposal costs or replace inefficient processes. Great advances are also being made in the conversion of biomaterials into traditional and novel chemical feedstocks. These topics are well represented in the pages of the journal *Green Chemistry* and by winners of the Presidential Green Chemistry Awards.

A second measure of progress in this realm is innovation: making completely new things or designing radically new ways of making known substances. New techniques in nanotechnology and DNA editing provide ever-more sophisticated methods for control at the molecular scale. We have only begun to explore the incredible variety of pathways and shortcuts generated as evolution drives organisms to operate with optimal energy and material efficiency, even as they make novel materials with unique functionalities. Chemists are only beginning to harness those methods. We can now manipulate individual atoms and molecules, and harness the exquisite precision of biological machinery to make materials of our own design. It remains to be seen if these powerful tools will be deployed with a green chemistry ethos.

Finally, measuring progress requires defining the metrics of success. Many people have recognized the challenge of identifying clear measures that allow easy comparisons and a comprehensive evaluation of sustainable practices in chemistry. The concept of percent yield was published as early as 1867 [8] and hasn't changed much since. In contrast to that simple measure of reaction efficiency, one single calculation cannot capture the many impacts defined in the 12 principles of green chemistry. Metrics are therefore still evolving, and several have been proposed. Dicks and Hent [9] recently summarized several common metrics. Of those, the E factor comes closest to a direct comparison with percent yield [10]. Jiménez-González et al. [11] compiled a selection of green metrics for the pharmaceutical industry, and many of which apply more broadly in chemistry. Some companies are developing their own metrics. For example, the French flavorings company, MANE, aims for a universally applicable score by assigning points to various individual metrics defined as "Green Motion" [12]. Looking beyond the laboratory, Jiménez-Gonzálex et al. [13] have developed a life cycle analysis tool for pharmaceutical syntheses. The authors admit that the life span considered is confined to sources of materials only through product production, rather than its ultimate fate. A full systems view would link the green chemistry of drug design and manufacture through distribution, use (and misuse), to the fate of these molecules in water treatment facilities and waste streams.

Economics certainly drives much progress in greening chemical manufacturing and use. However, economics also inhibits green improvements when they are seen as too expensive to implement. A challenge to move forward is to align greenness with economics by both decreasing the cost of greening and increasing the cost of employing unsustainable processes.

2.2 Outlook: Roadblocks to Progress

If so much green chemistry is already happening, why does progress seem so slow? Why is the revolution that I, as a graduate student, expected to be completed in a few years still unfolding at a glacial rate? Of course, my expectation of an instantaneous change in the 1990s says more about my naïveté at the time than about any fundamental problem with the approach. Nevertheless, even after the general recognition of green chemistry as an aspiration, roadblocks are apparent throughout the chemical enterprise. We can identify several. I classify them as (1) lack of knowledge or ability; (2) ends-justify-the-means thinking; (3) systems inertia; and (4) lack of financial or policy incentives.

The lack of knowledge or ability to undertake green chemistry reform can result from a lack of education about available methods or techniques. Most obviously, an enterprise may not have personnel who are knowledgeable enough to evaluate possible replacement solvents or catalysts. A related problem is uncertainty in selecting the appropriate metric of greenness, or lack of well-defined metrics that apply to the process in question. This issue is primarily one of education and outreach. Over the long term, it will be solved by embedding green chemistry into all chemistry education so that tomorrow's practitioners integrate it throughout. However, we don't have time to wait for the current workforce to retire. Our goal should be to empower practitioners to examine their processes, assess problem points, and identify and implement solutions. Many tools are becoming available to assist in greening various steps, for example, the EPA's solvent substitution program [14], and the green chemistry literature is growing rapidly. For large firms the savings in waste reduction undoubtedly offset the investment in green chemistry, but smaller companies and research laboratories may find it difficult to invest up-front efforts for later savings, indeed they may have few, if any, chemists on staff. One solution would be to deploy a Green Chemistry Extension Service, modeled on the U.S. Department of Agriculture's cooperative extension system. Such a program could train a cadre of green chemistry emissaries to visit, by invitation, enterprises engaged in chemical processes to assist practitioners in identifying key points of improvement and share successful methods.

A different type of knowledge gap, infinitely more challenging and interesting for researchers, is the set of problems for which green solutions do not yet exist. Here a true bottoms-up green chemistry approach needs to look not only at individual steps in a process, but consider whether the process is needed at all.

A second impediment to rapid conversion to greener techniques is ends-justify-the-means thinking. An example was the idea, expressed or not, in the pharmaceutical industry that "We're about saving lives, so we don't have the luxury to think about green chemistry." Similarly, academic researchers are guilty of feeling their work is too important or their time and funds are too limited to allow diversions into greening their laboratory procedures. By definition, traditional companies must prioritize profits as their ultimate objective; thus, they have a built-in end goal that can eclipse environmental means to reach it. That attitude has been changing and progress is visible in all these areas. In particular, the pharmaceutical world is demonstrating that green chemistry is fully compatible with their goals, most notably through the American Chemical Society Green Chemistry roundtable. Even the intense time-critical process of making radiolabeled pharmaceutical compounds can be transformed by green chemistry [15].

Common barriers to all kinds of change are systems inertia and historical decisions that "lock-in" operational steps. Simply stated, we are loath to tinker with methods that work. That hesitation is magnified for chemical processes, which often require many interconnected steps, each dependent on the others. Changing a solvent or a catalyst may require different equipment, a different separation method, a different suite of analytical methods subject to new interferences, and a new set of quality control procedures, as well as a reanalysis of safety considerations. Many of those steps seem auxiliary to the main purpose or product in question. Systems inertia applies to research laboratories as well as production sites. If my goal is to develop a new synthetic scheme, I don't want to divert energy to reinventing a greener analytical procedure for every step. Overcoming embedded systems inertia requires more than good intentions; it requires investments in time and effort that must be supported by incentives at an institutional level.

Systems inertia may also take the form of historically locked-in decisions, which present enormous barriers to change. The locked-in problem is made visible as physical infrastructure that is difficult and costly to change, but locked-in mind-sets may be equal to blame for inhibiting innovative solutions. Perhaps the biggest form of locked-in infrastructure is our commitment to fossil fuels as our primary energy supply. Coal drove the industrial revolution but it cannot fuel the twenty-first century. With the help of cheap and abundant fuel, chemical synthesis achieved dramatic gains over the past 150 years mainly through application of brute-force methods to break and recombine bonds through heat, pressure, and strong acids, bases, oxidants, and reductants. The traditional brown chemistry approach has been remarkably successful and has undergone many improvements since first being employed by early alchemists. However, as John Warner of Beyond Benign and others have pointed out, innumerable complex molecules are made every second by biological processes operating at ambient temperature and pressure in mostly aqueous environments. We should not allow the history of energy-intensive processing to lock-in our thinking and constrict our perceived universe of possible solutions. The fact that refluxing a mixture in an organic solvent has worked before does not mean it is the only method to effect a transformation. Indeed, the recent surge in reports of solvent-free synthetic procedures illustrates how reconsidering the simple notion, long considered gospel, that organic reactions proceed in organic solvents can be revolutionary [16].

Overcoming the impediments of knowledge or ability gaps, ends-versus-means myopia, and systems inertia requires a multipronged approach. Policies and incentives that are misaligned with sustainability goals contribute to all these roadblocks. At the foundation of each roadblock is the difficulty in balancing a narrow focus on a specific problem with a systems view that considers the global impacts of chemistry.

A comprehensive systems approach needs to account for the continuum of cumulative local actions that lead to planet-wide effects. Thus, policies and incentives must be coherently designed to influence small and large decision points across the enormous spectrum of chemistry activities. Incentives can act at the level of a single person in a single laboratory and extend up to the international scale of multinational corporations or multilateral treaties. A single researcher or laboratory team may select a target molecule to synthesize, a reaction solvent, or an analytical procedure, based on local and immediate incentives of design needs, cost, availability, or familiarity. Greening that process requires devising individual incentives for that person or laboratory. At the other extreme, a CEO's choice to invest (or not) in improving a global company's environmental footprint can be swayed by many factors, including, apparently, its executive compensation structure [17]. The impact on the global environment is the cumulative result of innumerable decisions, each of which may be nudged in a positive direction through both top-down and bottom-up pathways.

Historically, policies for managing the environmental impacts of chemicals have been seen as punitive. Strict rules were perhaps an inevitable first policy response to the widespread uncontrolled pollution of the early twentieth century. Obsolete attitudes that are still fixated on equating policy with constricting rules exemplify the concept of locked-in thinking. The idea that policies to improve human and environmental health always mean restrictive mandates on business is outdated and yet persists conspicuously in political rhetoric. In forward-looking business environments those attitudes are gradually going extinct, but their endurance over decades illustrates how difficult change can be. Of course, even as attitudes about successful policies evolve, existing policies are slow to change when they are locked into inflexible laws. Newer forward-thinking policies are flexible, adaptive, and iterative and include incentives as well as restrictions.

An example of a win–win business innovation is the emerging concept of "chemical leasing" promoted by the United Nations Industrial Development Organization [18]. The idea is that a company sells the service a chemical substance provides, rather than selling the substance itself, therefore, aligning the incentives of both provider and buyer to use chemical substances as efficiently as possible [19, 20]. These types of programs can be promoted by public policies. Ongoing assessment to understand which policies lead to the desired results is critical to their wide dissemination, adoption, and success.

Chemists in all types of institutions can personally promote the more rapid uptake of green chemistry. To return to the analogy with safety, in the pursuit of laboratory safety, many chemical journals have guidelines that require a description of hazards when they publish new chemical syntheses or procedures. Publishers could similarly support sustainability by demanding that authors report some key green metrics such as atom economy, types and amounts of waste generated, energy intensity, sources of feed stocks, and reasons for choosing solvents or materials that are less benign than typical. Initial steps in this direction would be easy: a description of a new synthesis is certainly expected to report percent yields, an extension to include atom economy would be trivial. Likewise, a listing of the waste materials produced is straightforward, while volume of waste per gram of product requires only slightly more effort. Similar steps should be designed and promoted for publications on other chemistry topics. Which analytical techniques are more environmentally benign? What standard green metrics should be reported for new catalysts?

Steps we can all immediately take as authors, reviewers, and student mentors is to continually raise questions about green chemistry and sustainability to keep these issues in the forefront. Authors who regularly receive reviews asking about the green aspects of their work will begin to incorporate those ideas into their papers, and more importantly, into their thought patterns. Students learn what is important by what is emphasized by their professors. Inclusion of green chemistry topics in class and on exams throughout the curriculum demonstrates their centrality to our field.

2.3 Summary

Moving the global economy to achieve net-zero carbon emissions requires a dramatic restructuring of the flow of energy and materials that are fundamental to the chemical enterprise. The goals of green chemistry are well aligned with those of traditional chemistry. While much progress has been made, obstacles remain to the full integration of sustainability into the chemical enterprise. This planetary imperative will employ scientists and engineers of all kinds for the foreseeable future. We recognize that solutions that lead to a sustainable world are based on chemistry. As contributors to both the problems and the solutions, and as inhabitants of the planet Earth, all chemists must be engaged in this project.

References

[1] Poliakoff, M.; Fitzpatrick, J. M.; Farren, T. R.; Anastas, P. T. Green Chemistry: Science and Politics of Change. *Science* 2002, *297* (5582), 807–810. DOI: 10.1126/science.297.5582.807.
[2] Woodhouse, E. J. Green Chemistry as Social Movement? *Sci Technol Hum Values* 2005, *30* (2), 199–222. DOI: 10.1177/0162243904271726.
[3] European Chemicals Agency. EC Inventory – ECHA. *echa.europa.eu*. November 7, 2015.
[4] Pena-Pereira, F.; Kloskowski, A.; Namieśnik, J. Perspectives on the Replacement of Harmful Organic Solvents in Analytical Methodologies: A Framework toward the Implementation of a Generation of Eco-Friendly Alternatives. *Green Chem* 2015, *17* (7), 3687–3705. DOI: 10.1039/C5GC00611B.
[5] Delidovich, I.; Palkovits, R. Catalytic versus Stoichiometric Reagents as a Key Concept for Green Chemistry. *Green Chem* 2016, *18* (3), 590–593. DOI: 10.1039/C5GC90070K.
[6] Anastas, P. T.; Kirchhoff, M. M.; Williamson, T. C. Catalysis as a Foundational Pillar of Green Chemistry. *Appl Catal A* 2001, *221* (1–2), 3–13. DOI: 10.1016/S0926-860X(01)00793-1.
[7] Velders, G. J. M.; Fahey, D. W.; Daniel, J. S.; Andersen, S. O.; McFarland, M. Future Atmospheric Abundances and Climate Forcings from Scenarios of Global and Regional Hydrofluorocarbon (HFC) Emissions. *Atmos Environ* 2015, *123* (Part A), 200–209. DOI: 10.1016/j.atmosenv.2015.10.071.
[8] Wright, C. R. A. On the Practical Losses in the Bleaching Powder Manufacture. *Chemical News* 1867, 126.
[9] P Dicks, A.; Hent, A. *Green Chemistry Metrics*. Springer International Publishing: London, 2015, pp. 1–95.
[10] Sheldon, R. A. E Factors, Green Chemistry and Catalysis: An Odyssey. *Chem Commun (Camb)* 2008, (29), 3352–3365. DOI: 10.1039/b803584a.
[11] Jiménez-González, C.; Constable, D. J. C.; Ponder, C. S. Evaluating the "Greenness" of Chemical Processes and Products in the Pharmaceutical Industry – A Green Metrics Primer. *Chem Soc Rev* 2012, *41* (4), 1485–1498. DOI: 10.1039/C1CS15215G.
[12] Phan, T. V. T.; Gallardo, C.; Mane, J. GREEN MOTION: A New and Easy to Use Green Chemistry Metric from Laboratories to Industry. *Green Chem* 2015, *17* (5), 2846–2852. DOI: 10.1039/C4GC02169J.

[13] Jiménez-González, C.; Ollech, C.; Pyrz, W.; Hughes, D.; Broxterman, Q. B.; Bhathela, N. Expanding the Boundaries: Developing a Streamlined Tool for Eco-Footprinting of Pharmaceuticals. *Org Process Res Dev* 2013, *17* (2), 239–246. DOI: 10.1021/op3003079.
[14] Harten, P. F. Program for Assisting the Replacement of Industrial Solvents (PARIS III); North Bethesda, MD. https://www.epa.gov/chemical-research/program-assisting-replacement-industrial-solvents-paris-iii Date accessed: August 26, 2016.
[15] Shao, X.; Fawaz, M. V.; Jang, K.; Scott, P. J. H. Ethanolic Carbon-11 Chemistry: The Introduction of Green Radiochemistry. *Appl Radiat Isot* 2014, *89*, 125–129. DOI: 10.1016/j.apradiso.2014.01.033.
[16] Tanaka, K.; Toda, F. Solvent-Free Organic Synthesis. *Chem Rev* 2000, *100* (3), 1025–1074. DOI: 10.1021/cr940089p.
[17] Minor, D. Executive Compensation and Misconduct: Environmental Harm. *SSRN J* 2016, 1–45. DOI: 10.2139/ssrn.2714438.
[18] United Nations Industrial Development Organization. Global Promotion and Implementation of Chemical Leasing Business Models in Industry. 2016, 1–60.
[19] Moser, F.; Jakl, T.; Joas, R.; Dondi, F. Chemical Leasing Business Models and Corporate Social Responsibility. *Env Sci Poll Res Int* 2014, *21* (21), 12445–12456. DOI: 10.1007/s11356-014-3126-y.
[20] Schwager, P.; Moser, F. The Application of Chemical Leasing Business Models in Mexico. *Env Sci Poll Res Int* 2006, *13* (2), 131–137. DOI: 10.1065/espr2006.02.294.

Heinz Plaumann

3 Switchable Polarity Solvents: Are They Green?

Abstract: Solvents play an incredibly important role in large scale chemical reactions. Switchable polarity solvents may prove to be a class of solvent that offers energy and material efficiencies greater than existing solvents. This paper examines such solvents and their potential in a variety of chemical reactions.

Keywords: Switchable polarity solvents

3.1 Introduction

In a monograph dedicated to green chemistry, it would be redundant and presumptuous for this author to add to an introduction proclaiming that green chemistry is indeed "good for us," good for the long-term sustainability, indeed survival, of our planet Earth. Switchable polarity solvents (SPS) do indeed align with many of the green chemistry principles [1].

The overall importance of solvents in the chemical process industry is paramount for separation and purification of countless materials, as well as reaction media. Their efficiency and effectiveness are often dependent upon their selectivity: solvation power and thermodynamic properties (volatility for recovery). The development of SPS allows chameleon-like behavior through judicious selection of physical conditions and through simple chemical reactions. We will show below that these solvents are no less "green" than Kermit himself!

3.2 Basic Chemistry: What Is an SPS?

The pioneering work of Jessop [2–7], his coworkers, and many collaborators defined the notion of SPSs illustrated best in their early work [4]: SWPs are liquids that can be converted easily and reversibly from one form to another, where the two forms differ significantly in hydrophilic/hydrophobic properties.

The simplified process is given in Figure 3.1. Consider an amine, with sufficient "alkyl" group power to be viewed in behavior as a nonpolar solvent. In this form, its behavior will nearly mimic that of an aliphatic hydrocarbon solvent with the expected solvation behavior dissolving "like" materials – oils, fats, and nonpolar organics. Reaction of said amine with CO_2 under controlled conditions will afford its conversion to the respective carbamic acid, now polar and with ionic solvation properties. The reaction is reversible to recover the original amine. CO_2 is the preferred trigger for the switching process: nontoxic, benign, inexpensive, and easily removed.

One can easily envisage that this may serve in extraction and separation processes, catalytic reactions requiring polar or nonpolar environments for favorable

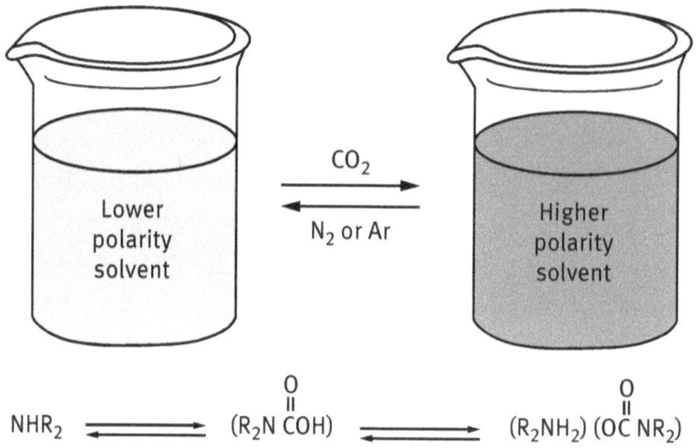

Figure 3.1: Switchable polarity solvent process showing conversion of the lower polarity solvent (e.g., aliphatic amine) in beaker on left to higher polarity carbamic acid and salt through reaction addition of carbon dioxide in beaker on right.

reaction control, for more "mundane" applications such as cleaning surfaces or substrates, control of coating behavior, and so on. This technology affords control of solution of the desired solute into its respective phase and subsequent recovery by switching the nature of the environment.

Subsequent treatment of this mixture with CO_2 under appropriate conditions (typically 1 atmosphere pressure) will convert the amine to the corresponding carbamic acid and then to the corresponding ionic carbamate salt. The solvent has been "switched" to ionic and highly polar, allowing its separation from the solute by extraction with water or small alcohol molecules. Reversion to the nonpolar form can be achieved through reduction of CO_2 pressure and application of heat.

An excellent overview, both descriptive and diagrammatic, is mentioned in Ref. [4] and also in Figure 3.1.

3.3 Process

Typically, conversion of the SPS from its nonpolar form to polar can be effected by simple addition of 1 bar CO_2 pressure. Reversion from its polar again to its nonpolar form is completed by bubbling argon or nitrogen through the medium. Of course, application to a commercially viable process requires further consideration of such factors as reaction vessel design, solvent separation of solvent whether in polar or nonpolar form, application of adequate CO_2 pressure, and supply of inert gas for reversion to nonpolar form.

Above all, the process steps for recovery of the target solute are most critical as are the overall economic considerations.

3.4 Application Examples

The versatility and potential of the SPS technology is in its infancy, with numerous illustrative examples given in the Patent of Ref. [9]. A sample listing of applications, proven to show commercial interest, if not yet success, is given below. It is the author's hope that, rather than an extensive survey of all applications, these examples will spur creative thinking to broaden their utility!

3.5 Extraction of Soybean Oil

In Canada alone, over 4,000 tons of hexane is emitted to the atmosphere [8] with about a third from oilseed processing. Recovery of such a solvent is also an energy-intensive process and carries many inherent safety concerns. Application of SPS is a nondistillative route for separation of the solvents from the products, without use of any volatile solvents, inherently safer and "greener" [4]. In simple laboratory tests, soy flakes were stirred with solvent, either hexane as control, or those amines under study, overnight. Mass ratio: 1:5 of soy flakes:solvent.

In this reference, the authors studied a range of amine-containing solvent for suitability and efficiency as SPS. For the work at hand, they selected amidines **1** and **2** for their desired phase behavior. Also, no efficiency for recovery of the soybean oil was given (8.8 mg extracted from 100 mg of flakes) (Figure 3.2).

3.6 Cleaning of Solid Particles

The removal of oil from solid particles can be carried out using SPS technology. Application of this to extract oil from tar sands and from oil-contaminated drilling fines is described in Refs [9, 10]. Conceivably the technology could be extended somehow to selective hard surface cleaning in a more usual detergent mode.

3.7 Recovery of Residual Motor Oil

Over 3 billion high density polyethylene (HDPE) containers are used annually for supplying motor oil to the consumer market [11]. The residual oil typically (2–5 % or

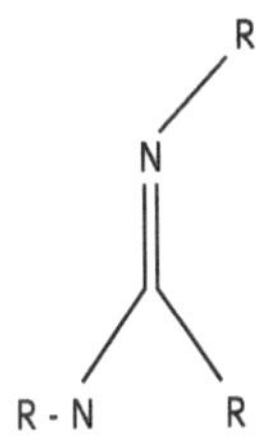

Legend

1: R = Propyl

2: R = Butyl

Figure 3.2: Propyl and butyl amidine analogues proven useful in the soybean extraction example.

20–50 mL) renders these containers unsuitable for recycling and these containers are merely disposed of to landfill. Simple rinsing of such containers or of their shredded plastic flakes [9] with the nonpolar version of the SPS would render the HDPE recyclable as well as allow for recovery of the oil upon conversion to the polar form and separation of the resulting biphasic mixture [12].

3.8 SPS as Reaction Medium

Conceptually, SPS can be viewed as a useful medium for organic reactions. The desired reaction could be performed in one form of the solvent after which it is converted to its other form to facilitate recovery of the product, catalyst, or other chemicals of interest. Reaction examples include Claisen–Schmidt condensations, Michael additions, cyanosilylations, and even air-sensitive Heck reactions (see Ref. [13] for a review). SPS technology has demonstrated utility in polymerization of styrene [14].

3.9 Recovery of Polystyrene from Polystyrene Foam

Polystyrene foam has an air content of about 90 % by volume [9, 15]. Transporting such low-density material for consolidation and recycling is very energy intensive (every color BUT green!) and the air in its own right can cause process problems. Recycling of such foam can only become the norm if an energy-efficient process of removing the air from the foam can be found and implemented.

SPS based on *N,N*-dimethyl cyclohexylamine was found to dissolve the polystyrene foam quickly, and the entrapped air was released as bubbles. Addition of this solution to carbonated water resulted in the rapid precipitation of the dense polystyrene plastic and caused the solvent to switch to its water-soluble ionic form. The solid polystyrene was separated and the solvent reverted to its hydrophobic form by slight heating and aeration to remove CO_2. Removal of the solvent by decantation rendered it ready for reuse.

3.10 Other Applications

A plethora of potential applications exist for SPS technology. For stimulus to innovative thinking, some are listed here: Improved deposition of dyes to textiles [9], purification of biodiesel [9], extraction of odorous compounds from plastic [9], catalyst recovery from a polymeric product [5], solvents for CO_2 capture [16], extraction of lipids and fatty acids from algae [17], switchable surfactants [16], and switchable solutes and catalysts [16, 19].

3.11 Future Considerations

As shown, SPS technology can be both energy and material efficient and effective. And as such, of considerable environmental advantage: That is, GREEN! The literature cited ranges from very precise chemical considerations and theoretical analyses to very practical processes. In particular, the extensive patent literature of not just the researchers but their many industrial collaborators underscores considerable commercial interest [18].

References

[1] Anastas, P. T.; Warner, J. C. *Green Chemistry: Theory and Practice*. Oxford University Press: New York, 1998. Also see www.acs.org/greenchemistry
[2] Jessop, P. G.; Hildebrant, D. J.; Xiaowang, L.; Eckert, C. A.; Liotta, C. L. Green chemistry: reversible nonpolar-to-polar solvent *Nature* 2005, *436*, 1102.
[3] Phan, L.; Andreattra, J.; Horvey, L. K.; Edie, C. E.; Luco, A.-L.; Mirchandani, A.; Darensbourg D. J.; Jessop, P. G. *J Org Chem* 2008, *73*, 127–132.
[4] Jessop, P. G.; Phan, L.; Carrier, A.; Robinson, S.; Duerr C.; Harjana, J. A solvent having switchable hydrophilicity *Green Chem* 2010, *12*, 809–814.
[5] Phan, L.; et al. (including P.G. Jessop). Switchable-polarity solvents prepared with a single liquid component *J Org Chem* 2008, *73*, 127–132.
[6] Boyd, A.; Jessop, P. G.; Dust J.; Buncel, E. Switchable polarity solvent (SPS) systems: probing solvatoswitching with a spiropyran (SP) - merocyanine (MC) photoswitch *Green Chem* 2013, *11*, 6047–6055.
[7] Durelle, J.; Vanderveen, J. R.; Quan, Y.; Chalifoux, C. B.; Kostin, J. E.; Jessop, P. G. Extending the range of switchable-hydrophobicity solvents *J Phys Chem Chem Phys* 2015, *17*, 5308–5313.
[8] National Pollutant Release Inventory, http://www.ec.gc.ca/inrpnrpi/ (accessed October 2009).
[9] Jessop, P. G.; et al., Switchable Hydrophilicity Solvents and Methods or Use Thereof. US Patent 89004444 B2, November 8, 2014 and patent references therein.
[10] Holland, A.; Wechsler, D.; Patel, A.; Molloy, B. M.; Boyd, A. R.; Jessop, P. G. Separation of Bitumen from Oil Sands using a Switchable Hydrophilicity Solvent *Can J Chem* 2012, *90*, 805–810.
[11] Wu, Q. Louisiana State University, Agricultural Center Report, Winter 2009.
[12] Jessop, P. G.; et al., 3rd International IUPAC Conference on Green Chemistry, August 15–18, 2010, Ottawa, Canada.
[13] Pollet, P.; Eckert, C. A.; Liotta, C. L. Switchable Solvents *Chem Sci* 2011, *2*, 609–614.
[14] Phan, L.; et al., Switchable solvents consisting of amidine/alcohol or guanidine/alcohol mixtures *Ind Eng Chem Res* 2008, *47*, 539–545.
[15] Jessop, P. G.; Kozycz, L.; Rahami, Z. G.; Schoenmakers, D.; Boyd, A.; Wechsler, D.; Holland, A. M. Tertiary amine solvents having switchable hydrophilicity *Green Chem* 2011, *13*, 619–623.
[16] Jessop, P. G.; Mercer, S. M.; Hildebrant, D. J. CO_2-triggered switchable solvents, surfactants, and other materials *Energy Environ Sci* 2012, *5*, 7240–7253.
[17] Samori, C.; et al., Extraction of hydrocarbons from microalga Botryococcus braunii with switchable solvents *Bioresour Technol* 2010, *101*, 3274–3279.
[18] Jessop, P.G. *Personal communication*, 2011.
[19] Jessop, P. G.; Eckert, C. A.; Liotta, C. L.; Heldebrant, D. J. US Pat. Appl. Publ. (2013) US 20130327989 A1 20131212, "Switchable Solvents and Methods of Use Thereof".

Martin Straka

4 Toward a Greenish Nuclear Fuel Cycle: Ionic Liquids as Solvents for Spent Nuclear Fuel Reprocessing and Other Decontamination Processes for Contaminated Metal Waste

Abstract: The final disposition of spent nuclear fuel (SNF) is an area that requires innovative solutions. The use of ionic liquids (ILs) has been examined as one means to remediate SNF in a variety of different chemical environments and with different chemical starting materials. The effectiveness of various ILs for SNF reprocessing, as well as the reaction chemistry that occurs in them, is discussed.

Keywords: ionic liquids, spent nuclear fuel reprocessing, decontamination, contaminated waste, actinides, lanthanides

4.1 Introduction

Dealing with spent nuclear fuel (SNF) and other radioactive waste management is one of the main issues of the nuclear industry. Compared with previous decades, the problem of radioactive waste is much more discussed by both public and professional communities, as the perception of the nuclear industry by the public is strongly affected by events like the Fukushima accident and it leads to increased demand for lower environmental affection and higher safety assurances. This is applied also to the radioactive waste management. Tighter regulation applied as a result by lawmakers adds an economic aspect into this. Sharply increasing costs for dealing with radioactive waste, altogether with pressure of public opinion, are strong motivation factors for development of recycling schemes, softening the effects on the environment, and bringing economic advantages for the nuclear industry. It should also be mentioned that a significant wave of decommissioning of nuclear facilities of the so-called first and second generations can be expected in the next decades and it makes the problem even more urgent. The Organisation for Economic Co-operation and Development/Nuclear Energy Agency estimates that 400 nuclear power plants will have to be decommissioned between 2014 and 2050 and it will generate more than 5 million tons of radioactive scrap metal. If the decommissioning of other nuclear installations is taken into account, almost 30 million tons of scrap metal contaminated by radionuclides will need to be handled [1]. Besides this, dealing with SNF still remains as a major problem of the nuclear industry. In both of these areas – SNF reprocessing and material decontamination – possible benefits of ionic liquids (ILs) are studied worldwide because of their interesting physical–chemical properties and chemical behavior.

For SNF reprocessing, ILs offer advantages of nonaqueous processes like a safer environment because of substituting volatile and flammable organic solvents or less serious criticality issues. At the same time, ILs beat high-temperature molten salt in comparing energy costs. For decontamination of radionuclide-contaminated materials, recycling is becoming strongly preferred over disposal. The motivation for development of recycling schemes is both social/environmental and economic. According to an International Atomic Energy Agency report [2], during the nuclear facility decommissioning process, only 3 % of material is considered contaminated, but costs for its disposal can make over 60 % of overall costs.

ILs (described as room-temperature molten salts in older literature) are commonly described as salts with organic cations associated with organic or inorganic anions that are liquid below 373 K and fully consisting of ions. In recent years, ILs gained an attention in many areas, including catalysis, separation processes, electrochemistry, or synthesis [3]. Another important area of ILs application is liquid–liquid extraction of actinides and fission products in case of liquid waste. Despite the fact that liquid–liquid extractions are also the problem that is important within the framework of recycling/decontamination or SNF reprocessing, it will not be covered here in detail as this chapter is focused only on solid waste. However, detailed reviews on this topic can be found in Refs [4–7]. Some of ILs' unique properties that are definitely attractive also for areas of SNF reprocessing and solid waste decontamination processes are reviewed later.

4.2 Radiation Stability of ILs

For any compound involved in the nuclear fuel cycle, radiation stability is a property of key importance as it will be exposed to high levels of radiation from actinides and fission products. Allen et al. [8] have shown that the resistance of 1,3-dialkylimidazolium nitrate and chloride to all forms of radiation is even significantly higher than that of the tri-*n*-butyl phosphate (TBP)/kerosene system widely used within PUREX reprocessing scheme (see later). Shkrob et al. have shown that ILs are not hypersensitive toward radiation in general [9].

Studies on radiation stability for other ILs are also available: 1-butyl-3-methylimidazolium hexafluorophosphate ([BMI][PF6], [10–12]), 1-butyl-3-methylimidazolium bis(trifluoromethylsulfonyl)amide ([BMI][NTf2], [10, 11, 13]), 1-butyl-3-methylimidazolium tetrafluoroborate ([BMI][BF4], [11, 12]), 1-butyl-3-methylimidazolium trifluoromethanesulfonate ([BMI][TfO], [11]) with the same results. Radiation stability of methyltributylammonium bis[(trifluoromethyl)sulfonyl]imide ([MBu3N][NTf2]) was studied by Bosse et al. [14]. Positive influence of introduction of FAP anion (tris (perfluoroalkyl)trifluorophosphate) was described by Guleria et al. [15], and it was suggested that this anion can shield the commonly used imidazolium cation showing the "designer" potential of ILs. Task-specific cations that include designated sites to assure deprotonation after radiolytic ionization were suggested by Shkrob and Marin [16]. Guanidium cation was found to be stable enough under radiation as well [17].

According to Huang et al. [18], radiation stability of 1-butyl-3-methylimidazolium chloride ([BMI][Cl]), which is one of the most widely used ILs, can be significantly improved by addition of $FeCl_3$ into the system.

Formation of gaseous products during radiolysis can represent a significant safety issue. This problem was addressed by Tarabek et al. [19] covering ILs based on $[NTf2]^-$ anion with several types of cations (ammonium, imidazolium, phosphonium, pyridinium, and pyrrolidinium).

Analysis of radiolytic stability of phosphonium-based ILs by Morco et al. [20] has shown relatively high resistance to radiolytic degradation, but highlighted a problem of small organic products of radiolysis which can, even in a small amount, agglomerate with IL to cause changes in the system properties. The same topic was covered also by Szreder et al. [21], who studied influence of aromatic ring ligands on the radiation stability of ammonium, imidazolium, and pyrrolidinium ILs. Results show that radiolytic degradation of studied ILs was significantly lower than that for *n*-alkanes in all cases [19].

A number of papers targeted radiation stability of more complex systems for extraction of several elements like Eu^{3+}, Sr^{2+}, and Cs^{+} or influence in the presence of HNO_3 [22–30]. Results show that radiolytic damage of ILs will not affect extraction performance of the system. ILs even protect the extractant as the IL is usually damaged instead of the extractant.

Results of the studies mentioned earlier suggest that ILs have very good radiation stability, which allows their use in nuclear applications. Moreover, in terms of radiation stability they are superior to organic solvents and reactants used today. This specific topic was critically reviewed by Mincher and Wishart [31].

4.3 Electrochemical Stability of ILs and Electrochemistry of Actinides and Lanthanides

Electrochemical deposition of actinides is an important step in high-temperature reprocessing schemes using both chloride and fluoride systems [32–34], and electrochemical separations from ILs are highlighted processes in the field of SNF reprocessing and other waste recycling schemes as well. As actinides and lanthanides usually react with water giving metal oxide/hydroxide and hydrogen, aqueous solutions cannot be used as carrier systems for electrodeposition process [35]. ILs, on the contrary, present a suitable environment for electrochemistry. Therefore, if electrochemical reduction steps will be included in any future SNF reprocessing or waste recycling flow sheets, as much data on electrochemical behavior of actinides and fission products as possible will be needed (see below). Knowledge of electrochemical behavior of pure ILs is equally important. Generally, ILs have good electrochemical stability (a wide electrochemical window). Characteristics of ILs for use as electrolytes can be found in Galiński et al. [36]. Based on data presented in this paper [36], a chart of potential windows for several ILs was prepared. As can be seen, potential windows are fully comparable with molten salt melts and can be as wide as 6 V (see Figure 4.1) for [BMI][BF4].

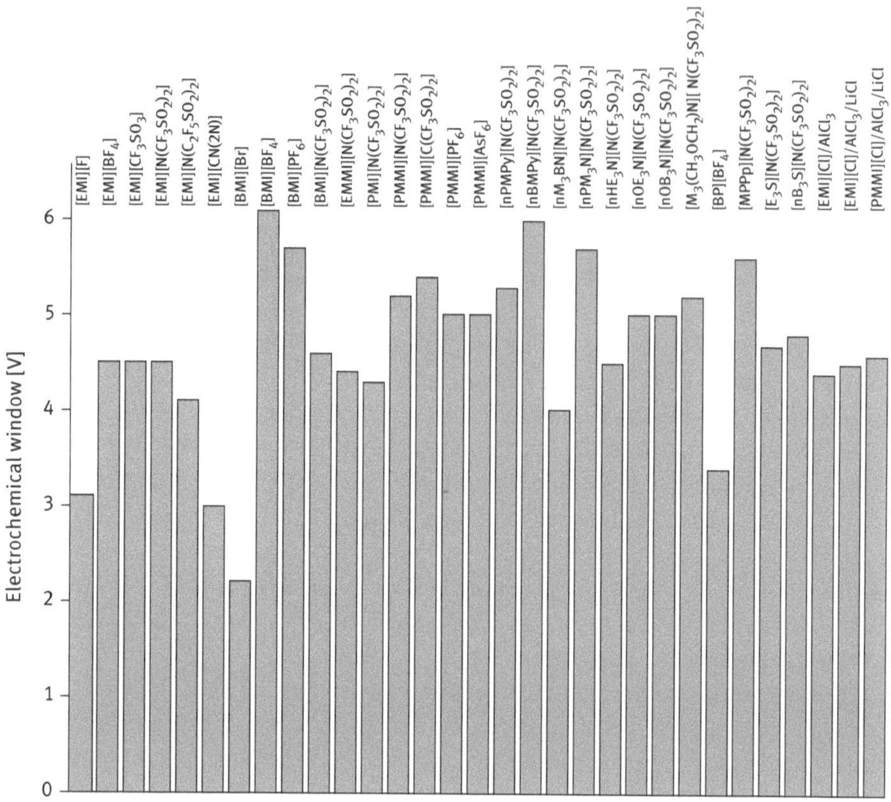

Figure 4.1: Potential windows of several ionic liquids based on data from Galiński et al. [36].

The electrochemistry of uranium was studied in several IL systems. The typical voltammetric scan of 1-ethyl-3-methylimidazolium chloride IL with U(VI) complexes can be seen in Figure 4.2. Two reduction peaks between −0.25 and −0.5 V can be attributed to U(VI) reduction to U(V) and subsequent reduction to U(IV), presumably in the form of UO_2. Most usually, uranium is in its tetravalent or hexavalent form at the beginning of an experiment. Nevertheless, stability of particular species as well as the mechanism of electrochemical reduction strongly depends on the system composition.

For example, electrochemistry of uranium in basic (in terms of Lewis acidity, $x_{(AlCl3)}/x_{(IL)} < 1$) $AlCl_3$ + *N*-(*n*-butyl)pyridinium chloride systems is described by Heerman et al. [38]. In this case, U(VI) is stable and the formation of $[UO_2Cl_{(4+x)}]^{(2+x)-}$ was described. U(VI) species readily underwent an irreversible reduction to UCl_6^{2-} according to the following equation:

$$[UO_2Cl_{(4-x)}]^{(2+x)-} + 2AlCl_4^- + 2e^- \rightarrow UCl_6^{2-} + 2AlOCl_2^- + (2+x)Cl^-.$$

U(III) was found to be unstable in basic melt and precipitated as solid UCl_3 [38].

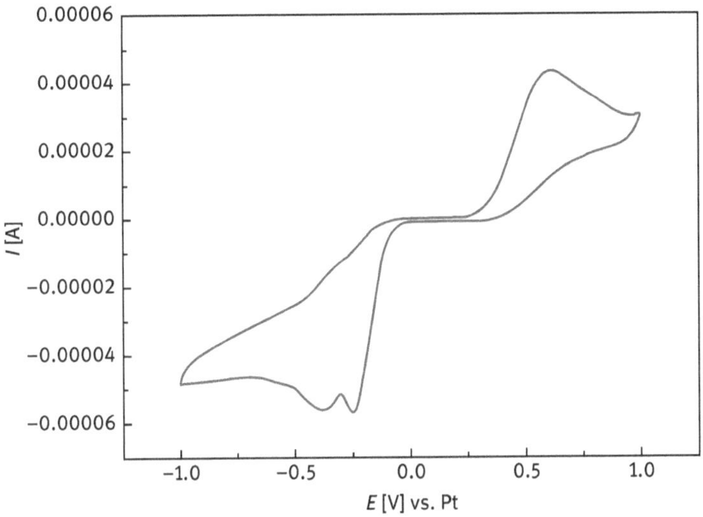

Figure 4.2: Cyclic voltammogram of the 1-ethyl-3-methylimidazolium/U(VI) system, three electrode setup, all electrodes made of Pt, scan rate 50 mV/s, $T = 353$ K [37].

On the other hand, in the acidic ($x_{(AlCl3)}/x_{(IL)} > 1$) $AlCl_3$/1-ethyl-3-methylimidazolium chloride system, color change from pale yellow to bright orange was observed within several days after addition of UO_2^{2+} [39], suggesting that UO_2^{2+} is converted to UCl_6^- by the reaction of the stating material with the IL. The mechanism of the conversion was suggested as follows:

$$UO_2^{2+} + Al_2Cl_7^- \leftrightarrow UO^{4+} + AlCl_4^- + AlOCl_3^{2-} \overset{Al_2Cl_7^{2-}}{\leftrightarrow} U^{6+} + AlCl_4^- + AlOCl_3^{2-}$$
$$U^{6+} + AlCl_4^- \leftrightarrow U^{5+} + AlCl_3 + \tfrac{1}{2}Cl_2.$$

It is in agreement with Dai et al. [40], who also described the stability of U(V) in terms of Lewis acidity dependence.

However, behavior of uranium can be dependent on the organic cation type as well. Nikitenko et al. [41, 42] studied spectroscopy and electrochemistry of U(IV) hexachloro complexes in BMINTf2 and Bu3MeNNTf2 ILs, and it was reported that electrochemical behavior of uranium species strongly depends on cation type. The mechanism of electrochemical reduction was the same in both ILs, but the potential of U(V)/U(IV) and U(IV)/U(III) redox pairs was shifted by 80 and 250 mV, respectively. This is because of different solvation of the uranium complexes in different ILs.

Successful experiment consisting of UO_2 dissolution and metallic uranium deposition was described by Jagadeeswara et al. [43]. Authors studied the electrochemical behavior of U(IV) in *N*-methyl-*N*-propyl-piperidinium bis(trifluoromethylsulfonyl)imide after the dissolution of UO_2 in the IL. The dissolution of UO_2 in

N-methyl-*N*-propyl-piperidinium bis(trifluoromethylsulfonyl)imide was quite rapid. The electrochemistry was studied at Pt, graphite carbon, and stainless steel working electrodes. Controlled potential electrolysis resulted in the deposition of metallic uranium [43].

In general, structure of IL affects the form in which uranium is present in the system as well as its mobility and the solvation environment in the system. All of these characteristics are very important for the electrochemical step in any of SNF reprocessing scheme. It can be also said that uranium can be directly deposited in the solid form and separated from the system, and the electrochemical reduction step is realistic. Except for experiments mentioned earlier, uranium reduction to solid form in different ILs has been thoroughly described [43–48].

Studies on electrochemistry of other actinides in ILs are very scarce compared to uranium. Electrochemical reduction of thorium in [BuMe3N][NTf2] was described by Bhatt et al. [49]. It was shown that thorium is reduced in one single step from Th(IV) to Th. Comparison of potentials for Th(IV)/Th pair with those obtained in various aqueous and high-temperature molten salt systems shows that thorium is reduced to its basic state much more easily in ILs than in other media. It can be attributed to the weak coordinating ability of the particular anion. Nevertheless, the layer of thorium oxide which is nonconductive was formed on the electrode surface and prevented further deposition of Th.

Nikitenko and Moisy [50] published their research of the electrochemical behavior of Pu and Np in [BMI][NTf2] IL. It was shown that both actinides form electrochemically inert complexes of $(AnCl_6)^{2-}$ type, and the reduction to An(III) was achieved only after the addition of chloride ions.

The data for electrochemistry of lanthanides – fission products – in ILs are also very rare. In aqueous systems, lanthanides occur mainly in the trivalent oxidation state (with the exceptions of europium and cerium). Other stable valencies are known mainly from solid state and nonaqueous systems [51]. In no contradiction with this statement, some ILs (especially acidic chloroaluminates) can stabilize several divalent lanthanide ions. Schoebrechts et al. [52, 53] reported successful electrolytic reduction of trivalent lanthanides to divalent ions (Sm^{2+}, Eu^{2+}, Tm^{2+}, and Yb^{2+}) in an $AlCl_3$/1-*n*-butylpyridinium chloride system. However, divalent lanthanides probably disproportionate according to the reaction suggested by O'Donnell [54], when acidity of a system is reduced:

$$3Ln^{2+} = 2Ln^{3+} + Ln.$$

Billard et al. [55] even reported that the stability of Eu^{2+} ions in [BMI][PF6] is so high that it cannot be oxidized by O_2, $K_2Cr_2O_7$, or $Ce(SO_4)_2$ and it is stable for months when left in the container.

In general, lanthanides (the same as actinides) are strongly electropositive and as a result, a quite stable carrier system is needed to achieve electrodeposition. As mentioned earlier, several ILs should be stable enough to provide a suitable environment.

Nevertheless, deposition experiments with lanthanides were reported very rarely so far. Al–La alloys were obtained from $AlCl_3$/EMIC system by Tsuda et al. [56]. Hsu and Yang were able to obtain cobalt together with dysprosium as a Co–Zn–Dy alloy from $ZnCl_2$/EMIC system [57]. Dysprosium was deposited also by Kurachi et al. [58] on a copper electrode from phosphonium bis(trifluoromethylsulfonyl)amide IL. Cyclic voltammograms of the system containing $Eu[N(SO_2CF_3)_2]_3$ in $[Me_4P][NTf_2]$ IL shows the Eu^{2+}/Eu^0 reduction step but no bulk electrolysis was done [59]. Ispas et al. [60] were able to deposit thin layers of samarium and cobalt/samarium from 1-butyl-1-pyrrolidinium bis(trifluoromethylsulfonyl)amide and were able to change deposit morphology by the parameters of electrolysis. Comparative experiments were done with quaternary ammonium ILs and 1-butyl-2,3-dimethylimidazolium triflate containing Y, Gd, and Yb ions [61]. It was shown that these rare earth elements can be deposited from ILs depending on electrode material and particular ILs.

It is clear from the previous text that some ILs have suitable electrochemical properties that allow us to consider them for usage in nuclear separation processes. On the other hand, the present level of knowledge is still insufficient and an extensive experimental campaign must be conducted to have a really detailed knowledge of basic behavior, which can then be applied for development of desired industrial processes.

4.4 Solubility of Actinides and Lanthanides in ILs

Prior to any separation of actinide or lanthanide species from IL systems, their initial form must be dissolved in it. ILs are quite often called "green solvents" which implies the good ability to dissolve a majority of materials. The truth is that solubility of inorganic compounds in common ILs is quite low [62]. For example, a solubility of $LaCl_3$ in [BMI][PF6] is only a 0.658 mg per 100 g of IL [63], and it takes more than 6 h to dissolve 0.1 g of UF_4 in 1 g of 1-butyl-3-methylimidazolium chloride (BMIC) even at 373 K [64]. The problem lies in the poor solvating power of ions like $[BF_4]^-$, $[PF_6]^-$, or $[NTf_2]^-$ and weak coordination of these ions. Of the most common ILs, those containing anions with better coordination provide quite good solubility of metal salts, including those of actinides and lanthanides. Chloroaluminate ILs can be taken as an example of such systems. In general, it is possible to increase solubility of particular compounds by designing a task-specific IL. One has to be also careful in interpreting experiments on solubility as sometimes even a small amount of water can solubilize the specie instead of IL and distort results. Also, air oxygen plays an important role in some cases. In Figure 4.3, one can see the UV-VIS spectra, proving the difference in uranium valence after dissolution of UF_4 in BMIC IL under different conditions as described by Ohashi et al. [64]. When the experiment is done under dry nitrogen atmosphere, uranium is then present in form of U(IV) ions. However, when the dissolution is done in normal atmosphere, air oxygen oxidizes U(IV) ions to U(VI) which supports the dissolution. Further study of solvation chemistry as well as development of new ILs designed to provide better solubility of relevant compounds is highly desirable.

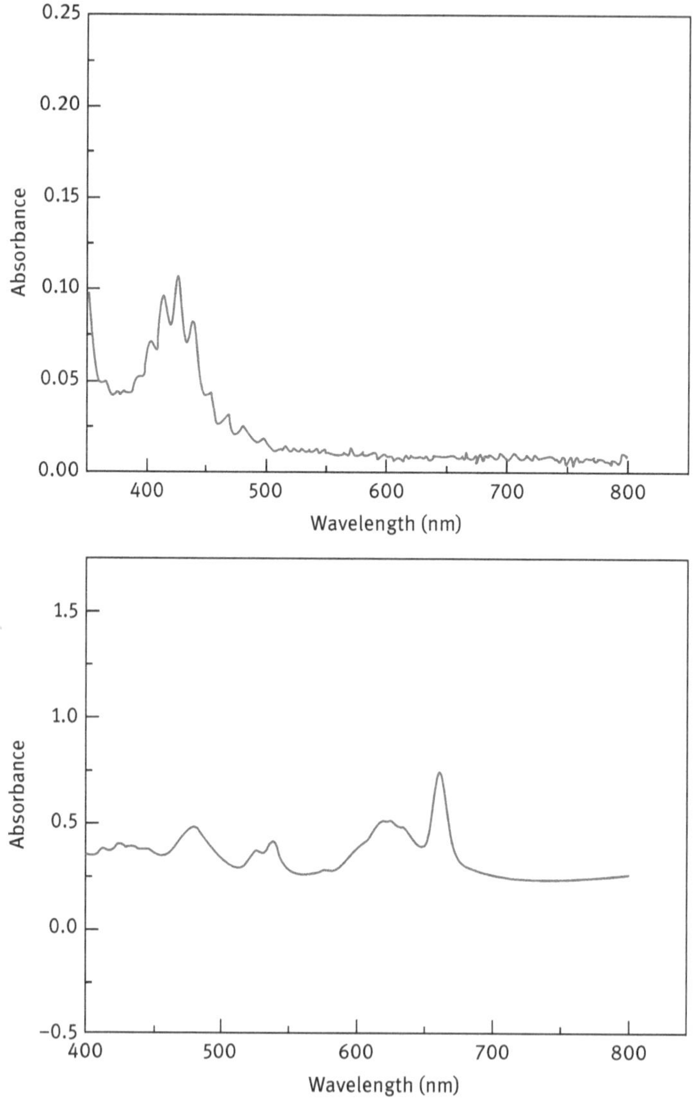

Figure 4.3: UV-VIS spectra of BMIC/UF4 system after dissolution period in normal atmosphere (left) and under dry nitrogen (right) [37].

4.5 ILs for Spent Fuel Reprocessing

Composition and processing of SNF depend on reactor type and on a fuel cycle applied. Throughout the world, the most common type of nuclear reactor is light water reactor, in which ^{235}U serves as a fissionable material. SNF consists mainly of uranium (^{235}U and ^{238}U) with a total of 95.6 %. Other actinides (Pu, Np, Am, and Cm) constitute about 1 % of SNF. Several groups of fission products containing elements like Cs, Sr, I, and

Tc take altogether 1 % of SNF [65]. Half-lives of these elements vary from several days to millions of years, which possess a threat to the environment and human health.

Two basic fuel cycle options can be described. A once-through fuel cycle counts with the period of short-lived fission products decay (SNF is placed in a water pond for up to 20 years) and then SNF is immobilized by vitrification and disposed of in a geological repository. This option eliminates the necessity of extensive difficult handling with material of very high level of activity. On the other hand, it requires a very robust geological repository for final disposal of the material that must be able to fulfill its duty for hundreds of thousands of years. On top of that, the major part of the material can be used again when properly recycled, and the once-through fuel cycle therefore causes also indirect environmental and economic impacts as the new raw material must be mined to replace the original one. On the other hand, a reprocessing fuel cycle contains steps that allow recovery of some elements for another use in the reactor. In its basic scheme, uranium and plutonium are recovered. For this, the so-called PUREX process is being used in an industrial scale. PUREX process is based on extraction of U and Pu by TBP in kerosene or *n*-dodecane [66]. Also, more advanced fuel cycles are considered to recover minor actinides and other fission products apart from U and Pu. Multiple use of SNF elements can significantly decrease the amount of highly active waste in terms of amount and activity that brings a lower environmental effect, lower risk of human health affection, and significant economic advantages (lower costs of waste disposal, no need for new raw material and its mining, multiple use or direct sale of decontaminated material, etc.).

Despite its industrial usage, the PUREX process is not suitable for many modern fuel types (inert matrix oxide, carbide, metallic, etc.) considered for the so-called Generation IV nuclear reactors [67, 68]. Therefore, interest in nonaqueous or pyrochemical reprocessing schemes is on the rise, with high-temperature molten salt technologies being the most prominent. For example, electrorefining reprocessing with uranium and transuranium elements recovery in an LiCl–KCl molten system is used by Argonne National Laboratory, USA, as a core process within the metallic fuel treatment flow sheet [69]. Robust high-temperature processes bring significant advantages for the reprocessing schemes with high radiation, thermal and electrochemical stability of molten salt systems, and high solubility of many compounds or an enhanced safety record including criticality and proliferation issues. On the other hand, at very high working temperatures, over 773 K, these processes can be very demanding of energy, corrosion resistance of used materials, or some particular safety issues connected to the temperature.

In recent years, ILs have drawn the attention of many research groups within the SNF reprocessing community. Usage of ILs instead of high-temperature molten salts for advanced SNF reprocessing can fulfill the requirements of most of the widely known 12 Principles of Green Chemistry as defined by Anastas and Warner [70]. Some of the most important intersections of ILs-based processes for SNF treatment and 12 Principles of Green Chemistry are described later.

Of course, the idea of SNF reprocessing as a whole strictly follows principle no. 1 – *to Prevent Waste*. ILs are commonly called “designer solvents” because of their tunability and wide options in synthesis to tackle specific problems [71]. This fact is therefore in agreement also with the fourth principle of green chemistry – *Designing Safer Chemicals*. Replacement of an organic fraction like kerosene in traditional extraction processes used for actinides and lanthanides is in accordance with the fifth principle of green chemistry – *Safer Solvents and Auxiliaries*. When comparing IL-based schemes with high-temperature molten salt processes, the sixth principle – *Design for Energy Efficiency* – is met because significant energy costs savings will be achieved. In proposed schemes for SNF reprocessing with usage of ILs, these are usually renewable input materials as well. For example, the electrochemical reduction can be done repeatedly in the same IL at least to some extent. Therefore, the seventh principle – *Use of Renewable Feedstock* – is fulfilled in that case. Two of the most highlighted characteristics of ILs are the negligible vapor pressure as well as nonflammability. It is fundamentally in agreement with the 12th principle – *Inherently Safer Chemistry for Accident Prevention* – because of much lesser risk of fire or explosion. For nuclear applications, criticality issues also belong under this principle and according to Harmon et al. [72], critical concentration of plutonium in Pu/IL mixtures are significantly higher than those in aqueous mixtures (a study was conducted for 1-ethyl-3-methylimidazolium tetrachloroaluminate, [EMI][$AlCl_4$], and 1-ethyl-3-methylimidazolium tetrafluoroborate, [EMI][BF_4], ILs), which brings another advantage to the process.

It should be mentioned that there are almost endless possibilities of ILs’ structure (according to Plechkova and Seddon [71], there are around 10^6 of possible simple IL systems, when binary and ternary systems are counted as well, 10^{12} and 10^{18} systems can be derived, respectively) and no statement such as those mentioned earlier can be understood as 100 % valid for every single system. There are known risks for some ILs such as production of toxic and highly corrosive hydrofluorine (HF) gas and other fluorine compounds during hydrolysis of ILs containing $[PF_6]^-$ anion [73]. Nevertheless, there is a significant intersection between the area of green chemistry and usage of ILs in general.

Baston et al. [74] suggested an alternative to molten salts electrorefining of SNF. High-temperature systems are substituted by an IL in this case. The scheme is based on anodic dissolution of SNF elements and selective electrodeposition of actinides in the second step. In Baston et al. [74], anodic dissolution of U and Pu metals was done in 1-ethyl-3-methylimidazolium chloride (EMIC) and the possibility of electrodeposition was verified by cyclic voltammetry. It should be noted that water can change the behavior of uranium when compared with chloroaluminate systems.

Asanuma et al. [75] proposed another simple scheme of SNF treatment process. In this scheme, two basic steps are realized in ILs: oxidative dissolution of SNF (which substitutes anodic dissolution from Ref. [74]) and electrochemical reduction with recovery of an UO_2 and UO_2/PuO_2 mixture, respectively (see Figure 4.4). In their previous

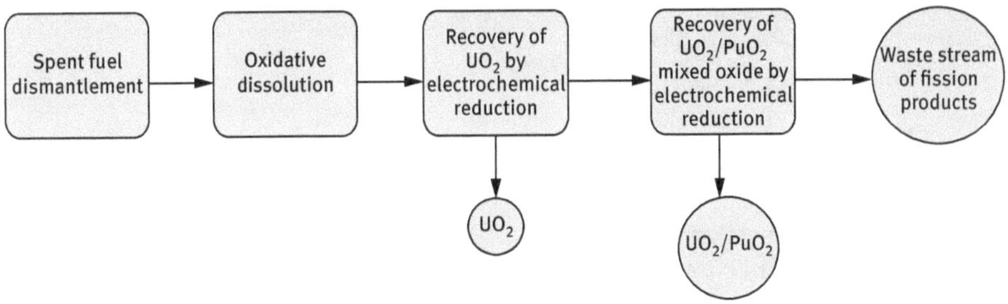

Figure 4.4: Scheme of SNF reprocessing using IL medium, based on Asanuma et al. [75].

papers, authors proved the successful oxidative dissolution of U_3O_8 by Cl_2 in 1-butyl-3-methylimidazolium nonafluorobutanesulfonate ([BMI][NfO]) as a demonstration of feasibility of the first step [76]. As a verification of feasibility of the electroreduction step, formation of the solid deposit after electroreduction was studied by Asanuma et al. [75]. Three IL systems were used: [BMI][Cl], [BMI][BF4], and [BMI][NfO]. Deposition of a mixture containing UO_2 and uranium oxychlorides was achieved in [BMI][NfO] [75].

4.6 Recycling/Decontamination Schemes

The same properties and characteristics of IL systems that were discussed within the SNF reprocessing can be used for decontamination/recycling schemes as well. Recycling will play an important role during future decommissioning of old nuclear facilities. Motivation for recycling was mentioned earlier. The general scheme of such a process can be seen in Figure 4.5. It can be seen that similarly to SNF reprocessing, it consists of two core processes. There is also an electrodeposition of radionuclides. In this case, it has the preceding step of washing of contaminated material during which the contaminant is transferred into the IL, and activity will decrease to the so-called clearance level.

As an example of a real application, steel elements contaminated with UF_4 were decontaminated by washing in BMIC IL to the clearance level, and subsequently, uranium was deposited on an electrode in the form of UO_2 and extracted from the IL [64]. UF_4 is the common contaminant of uranium in enrichment facilities or uranium refining and conversion plants, where it is formed by a reaction of gaseous UF_6 and Fe [77]. Deposit of uranium obtained from the contaminated material by Ohashi et al. [78] following the general scheme discussed earlier is shown in Figure 4.6 as well as the initial steel plate contaminated by uranium tetrafluoride. It is clear that the knowledge of electrochemical behavior of possible contaminants is a necessary knowledge before designing any process. This is a common thing with the spent fuel reprocessing area; and the state of the art in the electrochemistry of actinides and lanthanides in ILs was discussed previously. Apart from uranium, other common contaminants are, for example, europium or cobalt, which are the most common elements in a radionuclide inventory of a boiling water reactor prior to decommissioning [79].

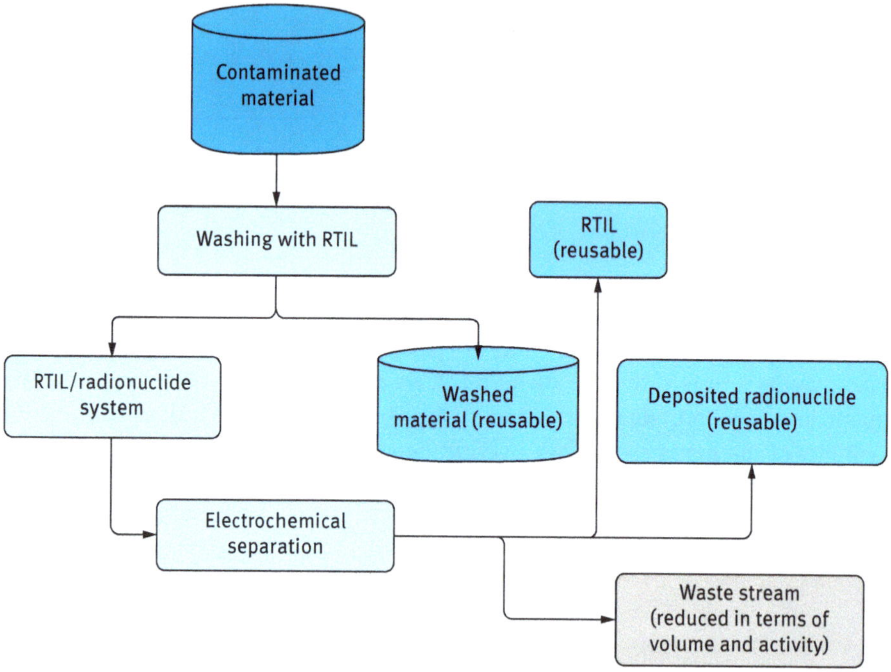

Figure 4.5: General scheme of the recycling process using ionic liquids. RTIL, room-temperature ionic liquid.

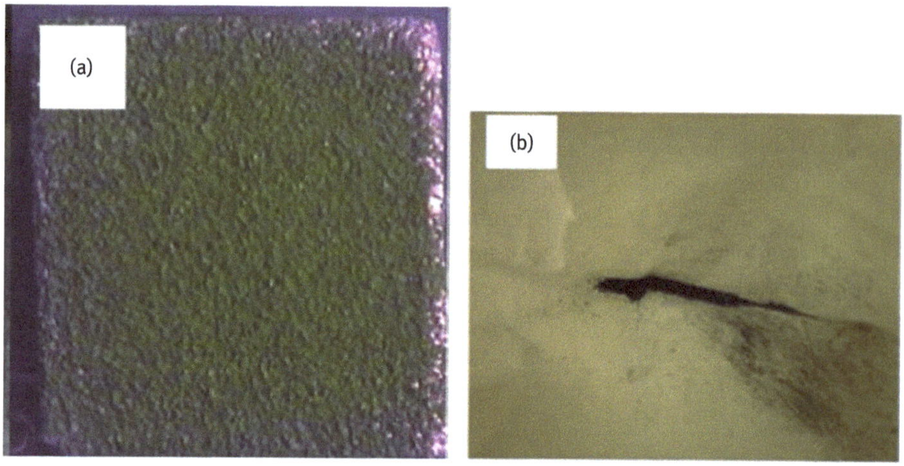

Figure 4.6: Steel plate contaminated by UF_4 and uranium deposit after washing of the material and electrochemical reduction.

Source: Pictures from Ohashi et al. [78].

4.7 Summary

In this chapter, several basic concepts of SNF or solid radioactive waste treatment using ILs as well as IL properties relevant for these schemes were summarized and presented. ILs surely have many unique characteristics, which can shift these demanding processes closer to the area defined by the 12 principles of green chemistry. Basic concepts were verified by several research groups but a lot of information on particular processes will need to be obtained in further studies. Detailed knowledge is needed in the field of solvation chemistry of actinides and lanthanides in ILs as well as in the field of electrochemistry of these elements in the same environment. However, the present lack of this knowledge doesn't mean that ILs should not be taken into account when thinking about the areas of SNF and radioactive waste because the added value of using these systems seems to be very promising. The nuclear sector has relatively low emission of carbon dioxide per unit of energy; therefore, it can play a very important role in combatting climate change. Concerns about the effect on the environment or human health by waste produced throughout whole fuel cycles are perfectly justified and it is necessary to study all possibilities that have potential ability to make all the processes safer and that lessen the effect on the environment. Processes based on ILs have the potential, but only further research will make it possible to reach an industrial scale.

References

[1] 7191 – R&D and Innovation Needs for Decommissioning Nuclear Facilities, 2014.

[2] *Managing Low Radioactivity Material from the Decommissioning of Nuclear Facilities*, Technical., no. 462. Vienna: International Atomic Energy Agency, 2008.

[3] Earle, M. J.; Seddon, K. R. Ionic Liquids. Green Solvents for the Future. *Pure Appl Chem* 2000, *72* (7), 1391–1398.

[4] Binnemans, K. Lanthanides and Actinides in Ionic Liquids. *Chem Rev* 2007, *107* (6), 2592–2614.

[5] Ha, S. H.; Menchavez, R. N.; Koo, Y.-M. Reprocessing of Spent Nuclear Waste Using Ionic Liquids. *Korean J Chem Eng* 2010, *27* (5), 1360–1365.

[6] Han, D.; Row, K. H. Recent Applications of Ionic Liquids in Separation Technology. *Molecules* 2010, *15* (4), 2405–2426.

[7] Pitner, W. R.; Bradley, A. E.; Rooney, D. W.; Sanders, D.; Seddon, K. R.; Thied, R. C.; Ratter, J. E. Ionic Liquids in the Nuclear Industry – Solutions for the Nuclear Fuel Cycle. in *Green Industrial Applications of Ionic Liquids*, NATO Science Series, Iraklion, Greece, 2003, 209–226.

[8] Allen, D.; Baston, G.; Bradley, A. E.; Gorman, T.; Haile, A. An Investigation of the Radiochemical Stability of Ionic Liquids. *Green Chem* 2002, *4*, 152–158.

[9] Shkrob, I.; Marin, T. W.; Wishart, J. F.; Grills, D. C. Radiation Stability of Cations in Ionic Liquids. 5. Task-Specific Ionic Liquids Consisting of Biocompatible Cations and the Puzzle of Radiation Hypersensitivity. *J Phys Chem B* 2014, *118* (35), 10477–10492.

[10] Berthon, L.; Nikitenko, S. I.; Bisel, I.; Berthon, C.; Faucon, M.; Saucerotte, B.; Zorz, N.; Moisy, P. Influence of Gamma Irradiation on Hydrophobic Room-Temperature Ionic Liquids [BuMeIm]PF6 and [BuMeIm](CF3SO2)2N. *Dalt Trans* 2006, *21*, 2526–2534.

[11] Le Rouzo, G.; Lamouroux, C.; Dauvois, V.; Dannoux, A.; Legand, S.; Durand, D.; Moisy, P.; Moutiers, G. Anion Effect on Radiochemical Stability of Room-Temperature Ionic Liquids Under Gamma Irradiation. *Dalt Trans* 2009, *31*, 6175–6184.

[12] Qi, M.; Wu, G.; Li, Q.; Luo, Y. γ-Radiation Effect on Ionic Liquid [bmim][BF4]. *Radiat Phys Chem* 2008, *77*, 877–883.
[13] Wang, S.; Liu, J.; Yuan, L.; Cui, Z.; Peng, J.; Li, J.; Zhai, M.; Liu, W. Towards Understanding the Color Change of 1-Butyl-3-Methylimidazolium Bis(trifluoromethylsulfonyl)imide During Gamma Irradiation: An Experimental and Theoretical Study. *Phys Chem Chem Phys* 2014, *16* (35), 18729–18735.
[14] Bossé, É.; Berthon, L.; Zorz, N.; Monget, J.; Berthon, C.; Bisel, I.; Legand, S.; Moisy, P. Stability of [MeBu3N][Tf2N] Under Gamma Irradiation. *Dalt Trans* 2008, *7*, 924–931.
[15] Guleria, A.; Singh, A. K.; Adhikari, S.; Sarkar, S. K. Radiation Induced Physicochemical Changes in FAP (fluoro alkyl phosphate) based Imidazolium Ionic Liquids and Their Mechanistic Pathways: Influence of Hydroxyl Group Functionalization of the Cation. *Dalt Trans* 2014, *43* (2), 609–625.
[16] Shkrob, I. A.; Marin, T. W. Radiation Stability of Cations in Ionic Liquids. 4. Task-Specific Antioxidant Cations for Nuclear Separations and Photolithography. *J Phys Chem B* 2013, *117* (47), 14797–14807.
[17] Shkrob, I.; Marin, T. W.; Bell, J. R.; Luo, H.; Dai, S. Radiation Stability of Cations in Ionic Liquids. 3. Guanidinium Cations. *J Phys Chem B* 2013, *117* (46), 14400–14407.
[18] Huang, L.; Huang, W.; Fu, H. Y.; Wu, G. Z.; Guo, Z. J.; Wu, W. S.; Chen, S. M. Influence of FeCl3 on Radiation Stability of Ionic liquid BmimCl. *Chinese Sci Bull* 2013, *58* (10), 1150–1155.
[19] Tarábek, P.; Liu, S.; Haygarth, K.; Bartels, D. M. Hydrogen Gas Yields in Irradiated Room-Temperature Ionic Liquids. *Radiat Phys Chem* 2009, *78* (3), 168–172.
[20] Morco, R. P.; Joseph, J. M.; Wren, J. C. The Chemical Stability of Phosphonium-Based Ionic Liquids Under Gamma Irradiation. *RSC Adv* 2015, *5* (36), 28570–28581.
[21] Szreder, T.; Skrzypczak, A. Influence of the Benzyl Substituent on Radiation Chemistry of Selected Ionic Liquids: Gaseous Products Analysis. *J Radioanal Nucl Chem* 2016, *307* (1), 195–202.
[22] Zhou, H.; Ao, Y.; Yuan, J.; Peng, J.; Li, J.; Zhai, M. Extraction Mechanism and γ-Radiation Effect on the Removal of Eu3+by a Novel BTPhen/[Cnmim][NTf2] System in the Presence of Nitric Acid. *RSC Adv* 2014, *4* (85), 45612–45618.
[23] Yuan, L.; Peng, J.; Xu, L.; Zhai, M.; Li, J.; Wei, G. Influence of γ-Radiation on the Ionic Liquid [C4mim][PF6] During Extraction of Strontium Ions. *Dalt Trans* 2008, *48*, 6358–6360.
[24] Yuan, L.; Peng, J.; Xu, L.; Zhai, M.; Li, J.; Wei, G. Radiation Effects on Hydrophobic Ionic Liquid [C4mim][NTf2] During Extraction of Strontium Ions. *J Phys Chem B* 2009, *113*, 8948.
[25] Yuan, L.; Xu, C.; Peng, J.; Xu, L.; Zhai, M.; Li, J.; Wei, G.; Shen, X. Identification of the Radiolytic Product of Hydrophobic Ionic Liquid [C4mim][NTf2] During Removal of Sr2+ from Aqueous Solution. *Dalt Trans* 2009, *38*, 7873–7875.
[26] Xu, C.; Yuan, L.; Shen, X.; Zhai, M. Efficient Removal of Caesium Ions from Aqueous Solution Using a Calix Crown Ether in Ionic Liquids: Mechanism and Radiation effect. *Dalt Trans* 2010, *39*, 3897–3902.
[27] Shkrob, I. A.; Chemerisov, S. D.; Wishart, J. F. The Initial Stages of Radiation Damage in Ionic Liquids and Ionic Liquid-Based Extraction Systems. *J Phys Chem B*, 2007, *111*, 11786.
[28] Shkrob, I. A.; Marin, T. W.; Chemerisov, S. D.; Wishart, J. F; Hatcher J. L. Radiation Induced Redox Reactions and Fragmentation of Constituent Ions in Ionic Liquids. 2. Imidazolium Cations. *J Phys Chem B* 2011, *115*, 3889.
[29] Shkrob, I. A.; Marin, T. W.; Chemerisov, S. D.; Wishart, F. Radiation and Radical Chemistry of NO3−, HNO3, and Dialkylphosphoric Acids in Room-Temperature Ionic Liquids. *J Phys Chem B* 2011, *115*, 3872.
[30] Shkrob, I. A.; Marin, T. W.; Chemerisov. S. D.; Wishart, J. F. No Title Radiation Induced Redox Reactions and Fragmentation of Constituent Ions in Ionic Liquids. 1. Anions. *J Phys Chem B* 2011, *115*, 3872.

[31] Mincher, B. J.; Wishart, J. F. The Radiation Chemistry of Ionic Liquids: A Review. *Solvent Extr Ion Exch* 2014, *32* (6), 563–583.

[32] Souček, P.; Malmbeck, R.; Mendes, E.; Nourry, C.; Glatz, J.-P. Exhaustive Electrolysis for Recovery of Actinides from Molten LiCl-KCl Using Solid Aluminium Cathodes. *J Radioanal Nucl Chem* 2010, *286* (3), 823–828.

[33] Serrano, K.; Taxil, P.; Dugne, O.; Bouvet, S.; Puech, E. Preparation of Uranium by Electrolysis in Chloride Melt. *J Nucl Mater* 2000, *282* (2–3), 137–145.

[34] Straka, M.; Korenko, M.; Lisý, F. Electrochemistry of Uranium in LiF–BeF2 Melt. *J Radioanal Nucl Chem* 2010, *284* (1), 245–252.

[35] Moeller, T. *The Chemistry of Lanthanides*. Chapman & Hall: London, 1965.

[36] M. Galiński, Lewandowski, A.; Stępniak, I. Ionic Liquids as Electrolytes. *Electrochim Acta* 2006, *51* (26), 5567–5580.

[37] Straka, M.; Szatmáry, L., Czech Chem. Soc. Symp. Ser. 2016, *14*, 159–310.

[38] Heerman, L.; DeWaele, R.; D'Olieslager, W. Electrochemistry and Spectroscopy of Uranium in Basic AlCl3 + N-(n-Butyl)Pyridinium Chloride Room Temperature Molten Salts. *J Electroanal Chem* 1985, *193*, 289.

[39] Anderson, C. J.; Choppin, G. R.; Pruett, D. J.; Costa, D.; Smith, W. Electrochemistry and Spectroscopy of UO22+ in Acidic AlCl3-EMIC. *Radiochim Acta* 1999, *84*, 31–36.

[40] Dai, S.; Toth, L.; Hayes, G.; Peterson, J. Spectroscopic Investigation of Effect of Lewis Basicity on the Valent-State Stability of an Uranium(V) Chloride Complex in Ambient Temperature Melts. *Inorg Chim Acta* 1997, *256* (1), 143–145.

[41] Nikitenko, S. I.; Cannes, C.; Naour, Le C.; Moisy, P.; Trubert, D. Spectroscopic and Electrochemical Studies of U(IV)-Hexachloro Complexes in Hydrophobic Room-Temperature Ionic Liquids [BuMeIm][Tf2N] and [MeBu3N][Tf2N]. *Inorg Chem* 2005, *44*, 9497.

[42] Nikitenko, S. I.; Hennig, C.; Grigoriev, M. S. Structural and Spectroscopic Studies of the Complex [BuMeIm]2[UCl6] in the Solid State and in Hydrophobic Room Temperature Ionic Liquid [BuMeIm][Tf2N]. *Polyhedron* 2007, *26* (13), 3136–3142.

[43] Jagadeeswara Rao, C.; Venkatesan, K. a.; Nagarajan, K.; Srinivasan, T. G.; Vasudeva Rao, P. R. Electrodeposition of Metallic Uranium at Near Ambient Conditions from Room Temperature Ionic Liquid. *J Nucl Mater* 2011, *408* (1), 25–29.

[44] Tan, X.-F.; Yuan, L.-Y.; Nie, C.-M.; Lui, K.; Chai, Z.-F.; Shi, W.-Q. Electrochemical Behavior of Uranyl in Ionic Liquid 1-Butyl-3-Methylimidazolium Chloride Mixture with Water. *J Radioanal Nucl Chem* 2014, *302* (1), 281–288.

[45] Giridhar, P.; Venkatesan, K. a.; Srinivasan, T. G.; Rao, P. R. V. Electrochemical Behavior of Uranium(VI) in 1-Butyl-3-Methylimidazolium Chloride and Thermal Characterization of Uranium Oxide Deposit. *Electrochim Acta* 2007, *52*, 3006–3012.

[46] Rao, C. J.; Venkatesan, K. a.; Nagarajan, K.; Srinivasan, T. G.; Rao, P. R. V. Treatment of Tissue Paper Containing Radioactive Waste and Electrochemical Recovery of Valuables Using Ionic Liquids. *Electrochim Acta* 2007, *53* (4), 1911–1919.

[47] Ikeda, Y.; Hiroe, K.; Asanuma, N.; Shirai, A. Electrochemical Studies on Uranyl(VI) Chloride Complexes in Ionic Liquid, 1-Butyl-3-Methylimidazolium Chloride. *J Nucl Sci Technol* 2009, *46*, 158–162.

[48] Sengupta, A.; Murali, M. S.; Mohapatra, P. K. Role of Alkyl Substituent in Room Temperature Ionic Liquid on the Electrochemical Behavior of Uranium Ion and Its Local Environment. *J Radioanal Nucl Chem* 2013, *298* (1), 209–217.

[49] Bhatt, A. I.; Duffy, N. W.; D. Collison, May, I.; Lewin, R. G. Cyclic Voltammetry of Th(IV) in the Room-Temperature Ionic Liquid [Me3NnBu][N(SO2CF3)2]. *Inorg Chem* 2006, *45* (4), 1677–1682.

[50] Nikitenko, S. I.; Moisy, P. Formation of Higher Chloride Complexes of Np(IV) and Pu(IV) in Water-Stable Room-Temperature Ionic Liquid [BuMeIm][Tf 2 N]. *Inorg Chem* 2006, *45* (3), 1235–1242.

[51] Evans, W. J. Perspectives in Reductive Lanthanide Chemistry. *Coord Chem Rev* 2000, *206–207*, 263–283.
[52] Schoebrechts, J. P.; Gilbert, B. P.; Duyckaerts, G. Electrochemical and Spectroscopic Studies of the Lanthanides in the AlCl3+1-n-Butylpyridinium Chloride Melt at 40 °C: Part II. The Tm(III-II), Eu(III-II) Systems, Application of Nugent's Linearization Method. *J Electroanal Chem* 1983, *145*, 139.
[53] Schoebrechts, J. P.; Gilbert, B. P.; Duyckaerts, G. Electrochemical and Spectroscopic Studies of the Lanthanides in the AlCl3+1-n-Butylpyridinium Chloride Melt at 40 °C: Part I. The Yb(III-II), Sm(III-II) Systems. *J Electroanal Chem* 1983, *145*, 127, 1983.
[54] O'Donell, T. A. *Superacids and Acidic Melts as Inorganic Chemical Reaction Media*. VCH Verlagsgesellschaft m.b.H.: Weinheim, 1993.
[55] Billard, I.; Moutiers, G.; Labet, A.; A. El Azzi, Gaillard, C.; Mariet, C.; Lützenkirchen, K. Stability of Divalent Europium in an Ionic Liquid: Spectroscopic Investigations in 1-Methyl-3-Butylimidazolium Hexafluorophosphate. *Inorg Chem* 2003, *42* (5), 1726–1733.
[56] Tsuda, T.; Nohira, T.; Ito, Y. Electrodeposition of Lanthanum in Lanthanum Chloride Saturated AlCl3-1-Ethyl-3-Methylimidazolium Chloride Molten Salts. *Electrochim Acta* 2001, *46* (12), 1891–1897.
[57] Hsu, H. Y.; Yang, C. C. Conductivity, Electrodeposition and Magnetic Property of Cobalt(II) and Dysprosium Chloride in Zinc Chloride-1-Ethyl-3-Methylimidazolium Chloride Room Temperature Molten Salt. *Zeitschr Nat Sect B J Chem Sci* 2003, *58* (2–3), 139–146.
[58] Kurachi, A.; Matsumiya, M.; Tsunashima, K.; Kodama, S. Electrochemical Behavior and Electrodeposition of Dysprosium in Ionic Liquids Based on Phosphonium Cations. *J Appl Electrochem* 2012, *42* (11), 961–968.
[59] Bhatt, A. I.; May, I.; Volkovich, V. A.; Hetherington, M. E.; Lewin, B.; Thied, R. C.; Ertok, N. Group 15 Quaternary Alkyl Bistriflimides: Ionic Liquids with Potential Application in Electropositive Metal Deposition and as Supporting Electrolytes. *J Chem Soc Dalt Trans* 2002, *24*, 4532–4534.
[60] Ispas, A.; Buschbeck, M.; Pitula, S.; Mudring, A.; Uhlemann, M.; Bunde, A.; Endresf, F. Electrodeposition of Co, Sm and Co-Sm Thin Layers. *ECS Trans* 2009, *16* (45), 119–127.
[61] Glukhov, L. M.; Greish, A. A.; Kustov, L. M. Electrodeposition of Rare Earth Metals Y, Gd, Yb in Ionic Liquids. *Russ J Phys Chem A* 2010, *84* (1), 104–108.
[62] Marciniak, A. The Solubility Parameters of Ionic Liquids. *Int J Mol Sci* 2010, *11* (5), 1973–1990.
[63] Branco, L. C.; Rosa, J. N.; Moura Ramos, J. J.; Afonso, C. a. M. Preparation and Characterization of New Room Temperature Ionic Liquids. *Chem – A Eur J* 2002, *8* (16), 3671–3677.
[64] Ohashi, Y.; Asanuma, N.; Harada, M.; Wada, Y.; Matsubara, T.; Ikeda, Y. Application of Ionic Liquid as a Medium for Treating Waste Contaminated with UF4. *J Nucl Sci Technol* 2009, *46* (8), 771–775.
[65] Nagy, S. (ed.), *Radiochemistry and Nuclear Chemistry*. EOLSS Publishers Co., Ltd., Oxford, 2009.
[66] Glatz, J.-P.; Spent Fuel Dissolution and Reprocessing Processes, in *Comprehensive Nuclear Materials*, Konings, R. (ed.), Elsevier Ltd, Amsterdam, 2012, 343–366.
[67] A Technology Roadmap for Generation IV Nuclear Energy Systems. 2002.
[68] Bourg, S.; Peron, F.; Lacquement, J. The Evaluation of the Pyrochemistry for the Treatment of Gen IV Nuclear Fuels – Inert Matrix Chlorination Studies in the Gas Phase or Molten Chloride Salts. *J Nucl Mater* 2007, *360* (1 Spec. Iss), 58–63, 2007.
[69] Accelerator-Driven Systems (ADS) and Fast Reactors (FR) in Advanced Nuclear Fuel Cycles – A Comparative Study. 2002.
[70] Anastas, P. T.; Warner, J. C. *Green Chemistry: Theory and Practice*. Oxford University Press: New York, 1998.
[71] Plechkova, N. V.; Seddon, K. R. Applications of Ionic Liquids in the Chemical Industry. *Chem Soc Rev* 2008, *37* (1), 123–150.

[72] Harmon, C. D.; Smith, W. H.; Costa, D. a. Criticality Calculations for Plutonium Metal at Room Temperature in Ionic Liquid Solutions. *Radiat Phys Chem* 2001, *60*, 157–159.

[73] Huddleston, J.; Visser, A.; Reichert, W.; Willauer, H.; Broker, G.; Rogers, R. Characterization and Comparison of Hydrophilic and Hydrophobic Room Temperature Ionic Liquids Incorporating the Imidazolium Cation. *Green Chem* 2001, *3* (4), 156–164.

[74] Baston, G. M. N.; Bradley, A. E.; Gorman, T.; Hamblett, I.; Hardacre, C.; Hatter, J. E.; Healy, M. J. F.; Hodgson, B.; Lovell, K. V.; Newton, G. W. A.; Nieuwenhuyzen, M.; Pitner, W. R.; Rooney, D. W.; Sanders, D.; Seddon, K. R.; Thied, R. C. Ionic Liquids for the Nuclear Industry: A Radiochemical, Structural, and Electrochemical Investigation. *ACS Symp Ser* 2002, *818*, 162–177.

[75] Asanuma, N.; Harada, M.; Yasuike, Y.; Nogami, M.; Suzuki, K.; Ikeda, Y. Electrochemical Properties of Uranyl Ion in Ionic Liquids as Media for Pyrochemical Reprocessing. *J Nucl Sci Technol* 2007, *44* (3), 368–372.

[76] Ikeda, Y.; Asanuma, N.; Harada, M. Application of Ionic Liquids to Pyrochemical Reprocessing Methods (1) – Dissolution of Uranium Oxide Fuels by Using Cl2. in *2005 Fall Meeting of the Atomic Energy Society of Japan*, 2005, J39.

[77] Dixmier, J.; Hasson, R.; Maraval, S.; Vincent, L. M. Corrosion Par L' Hexafluorure D ' Uranium. *J Nucl Chem* 1961, *3* (1), 41–59.

[78] Ohashi, Y.; Harada, M.; Asanuma, N.; Ikeda, Y. Feasibility Studies on Electrochemical Recovery of Uranium from Solid Wastes Contaminated with uranium Using 1-Butyl-3-Methylimidazorium Chloride as an Electrolyte. *J Nucl Mater* 2015, *464*, 119–127.

[79] IAEA. Radiological Characterization of Shut Down Nuclear Reactors for Decommissioning Purposes. 1998.

Lesley Putman, Chris Nyland and Kristine Parson

5 Green Disposal of Waste Bisphenol A

Abstract: Bisphenol A (BPA) is an important precursor for polycarbonates and epoxy resins that are used to make products that many would consider essential for modern living. Unfortunately, BPA is an endocrine disrupter and more and more evidence is being uncovered about its negative effects on humans as well as on organisms in the environment. Sustainable chemistry means we take everything into account in designing a chemical process: the risks, the costs, the benefits, and so on. BPA is not likely to be replaced by a less toxic chemical in the foreseeable future, nor is our society willing to give up valuable polycarbonate products. As part of sustainable green chemistry, we want to assure that any toxic chemicals that leave a manufacturing plant pose a minimal risk to the environment. Using a green method to clean up effluent before it is released into the environment would be ideal. In this report, little bluestem seeds are shown to be capable of degrading BPA in aqueous solutions. This gives them the potential to be used in reducing the amount of BPA that may be in effluent. Additionally, an enzyme was isolated that was responsible for degrading BPA and this could also have potential for the treatment of effluent.

Keywords: Bisphenol A (BPA), little bluestem, green chemistry, sustainable chemistry, seed exudate, germination, enzymes, degradation

5.1 Introduction

Bisphenol A (BPA) is an important industrial chemical that is used in the manufacture of epoxy resins and polycarbonates. According to the U.S. Environmental Protection Agency, BPA is considered a high-production volume chemical [1]. Merchant Research & Consulting published a report stating that the production of BPA in 2011 was approximately 4.4 million tons, increasing to 4.6 million tons in 2012, and predicted to be greater than 5.4 million tons by 2015 [2]. In order to make epoxy resins, BPA is combined with epichlorohydrin to make the monomer unit, then the resin is cured by reacting it with other chemicals to impart specific properties. Epoxy resin is used in making water pipes, flooring, wind rotor blades, marine coatings, automotive coatings, liners of food containers, fiber optics, and circuit boards. The other major use of BPA is in making the monomeric unit of polycarbonate plastics. Polycarbonate plastics are shatter resistant, are heat resistant, are lightweight, and have optical clarity. These properties have led to their use in safety equipment (helmets and visors), eye glasses, medical equipment, car components (headlamps and sunroofs), and coatings on CDs and DVDs. As of 2013, there was an annual production of 3.3 million tons of polycarbonate produced worldwide [3].

Epoxy resins and polycarbonates are an integral part of our daily lives and are not likely to be replaced with other materials in the foreseeable future. With concerns about BPA leaching out of products and having negative health effects, some have tried to green up the process of manufacturing plastics by looking for substitutes for BPA. For example, BPS (4,4′-sulfonyldiphenol) and BPF (4,4′-dihydroxydiphenylmethane) have both been used as alternatives to BPA but they also show endocrine disrupting effects like BPA [4, 5]. A novel substitute for BPA was proposed by Reno et al. [6] at an American Chemical Society meeting in 2014. They synthesized bisguaiacol F (BGF) from vanilyl alcohol and guaiacol which are lignin components. Polymers made with BGF had similar properties to those made with BPA, but rigorous testing of the polymers and of the health effects of BGF have not been completed. For now, it appears that the best approach is to try to keep BPA out of the environment and determine how to minimize human exposure to BPA.

Although landfills and household waste contribute to BPA in the environment, the largest contributor of BPA to the environment is industrial waste [1]. In making epoxy resin, BPA can be released when the monomer unit for epoxy resin is washed [7]. It is also released in the manufacture of products made from epoxy resin. The Epoxy Resin Committee commissioned an independent agency to determine potential emissions of BPA from the manufacture and use of epoxy resins [7]. They reported that in the manufacturing of marine coatings there is a maximum annual release of 194 kg of BPA. Maximum annual BPA releases for other manufacturing processes include 171 kg for flooring, 105 kg for automotive coating, 92 kg for wind rotor blades, and 11 kg for water pipes. The manufacturing of polycarbonates also produces BPA waste. In Nigeria, the effluents from plastic companies contained 107–162 μg/L BPA [8].

Although BPA has a relatively short half-life in the environment (4.5–4.7 days) [9] due to degradation by microorganisms [10, 11], there is a continual release of it into the environment giving a persistent detectable concentration. In sampling streams throughout the United States, BPA was found at a maximum concentration of 12 μg/L and the median detectable concentration was 0.14 μg/L [12]. There is no agreement on a minimum concentration of BPA which is safe for wildlife because toxic effects have been seen at very low levels for some organisms at specific stages of their development, while others require higher levels of BPA to see any effects. For example, when the crustacean, *Tigriopus japonicus* was exposed to 0.1 μg/L BPA for 4 weeks, its development was inhibited [13]. When the Western clawed frog (*Silurana tropicalis*) was exposed to 2.28 μg/L BPA for 9 days, spontaneous metamorphosis was inhibited [14]. Sufficient data regarding the toxicity of BPA to various organisms in the environment has accumulated [15] to convince Canada to require manufacturers to develop plans to deal with BPA in their waste, and propose a limit of 1.75 μg/L for emissions by manufacturers [16]. Environmental effects of BPA and mechanisms of action are outlined in a paper by Canesi and Fabbri [17]. They showed that BPA can cause developmental and reproductive effects in certain species of aquatic vertebrates; however, additional studies are necessary to establish dose–response curves and a more accurate picture of how aquatic organisms respond to BPA over the course of their life cycle.

Methods to remove BPA from wastewater include photodegradation using UV radiation with microwave and heat [18], biodegradation by white rot fungi and enzymes isolated from these fungi [19], and biodegradation by bacteria [20]. Plants can also degrade BPA. For example, cultured suspension cells of eucalyptus were shown to degrade BPA [21] and cultured suspension cells of tobacco as well as tobacco seedlings were found to take up and degrade BPA [22]. *Portulaca oleracea* was found to remove nearly all of 50 μM BPA from water in 24 h [23]. Water convolvulus, an aquatic plant, removed BPA from an aqueous medium and in 1 week the BPA was completely metabolized [24]. *Dracaena sanderiana*, a tropical evergreen plant, was able to take up BPA and was subsequently shown to be effective in removing BPA from landfill leachate [25].

In this study, we will be looking at little bluestem seeds (*Schizachyrium scoparium*) and their ability to degrade BPA as they germinate. Little bluestem is a perennial grass native to the United States and Canada and is used for prairie restoration [26]. It has also been used in phytoremediation studies for the degradation of petroleum hydrocarbons [27, 28]. This concept of seeds degrading chemicals as they germinate was first reported by Loffredo et al. [29] who used seeds of various plants and found that a number of them were able to degrade BPA as they were germinating.

5.2 Materials and Methods

5.2.1 Chemicals

BPA was obtained from Alfa Aesar, bis(4-hydroxyphenyl)methane (bisphenol F) was obtained from Aldrich, and BSTFA (N,O-bis(trimethylsilyl)trifluoroacetamide) was obtained from Supelco. All other chemicals were reagent grade except for methanol which was HPLC grade. Little bluestem seeds were obtained from Outside Pride (Oregon).

5.2.2 Degradation/Uptake of BPA by Little Bluestem Seeds

Little bluestem seeds [10] were added to 6 mL of 46 mg/L BPA in small (60 × 15 mm) Petri dishes. Control plates contained 6 mL of BPA solution (46 mg/L) but no seeds. All experiments were run in quadruplicate. Plates were kept in the dark at 27 °C. After the incubation time, 1 mL of solution was removed, internal standard (BPF) was added, then the solution was extracted with 1 mL of ethyl acetate to recover BPA and BPF. The ethyl acetate was evaporated and the residue was derivatized for gas chromatography/mass spectrometry (GC/MS). Seeds were removed from each plate and extracted with methanol containing internal standard (1 mL methanol for 10 seeds) by shaking in a test tube for 10 min at 200 rpm. Methanol was decanted from seeds and evaporated before derivatization for GC-MS.

5.2.3 Quantification of BPA

Residue containing BPA was derivatized by adding 50 μL each of BSTFA and pyridine, as described by Gatidou et al. [30], then the derivatized BPA was quantified by GC-MS. The instrument was a Finnegan Trace GC Ultra with Trace DSQ and AS 3000 autosampler (Thermo Electron Corp.). The column was a Phenomenex Zebron Inferno, ZB-5HT, 30 m × 0.25 mm i.d. and 0.25 μm. The inlet and the mass spec transfer line were at 280 °C. Injection was splitless and the carrier gas (He) flowed at 1.2 mL/min. The gas chromatograph program was 150 °C for 1 min, then the temperature was increased to 225 °C at a rate of 35 °C/min and held for 2 min. Finally, the temperature was increased to 325 °C at a rate of 25°/min and held there for 3 min. Detection mode was SIM, detecting positive ions with m/z of 179, 329, 344, 357, and 372.

5.2.4 Degradation of BPA by Seed Exudate

Seed exudate was prepared by soaking little bluestem seeds in deionized water in the dark at 27 °C for 6–10 days. After the incubation time, the seeds were removed and the remaining solution was the seed exudate.

Exudate solution (0.5 mL) was added to 0.5 mL of 0.100 M potassium phosphate pH 6.5 containing BPA giving a final concentration of 30 mg/L BPA in the incubation mixture. Controls contained deionized water instead of exudate solution. All were done in quadruplicate. Solutions were incubated at 27 °C for 4–8 days. After the incubation time, internal standard (BPF) was added, then the solution was extracted with 1 mL of ethyl acetate as mentioned earlier and derivatized for GC/MS analysis.

5.2.5 Sephadex G-75

Freeze-dried exudate (3.7 mg) was applied to a Sephadex G-75 column (34 × 2 cm). The column was eluted with 0.100 M potassium phosphate pH 6.5 and 1 mL fractions were collected. Absorbance was measured at 280 nm. Fractions obtained from the column were dialyzed against water and freeze-dried before being tested for their activity to degrade BPA. Blue dextran and DNP-aspartate were the standards used to determine the void volume and total volume of the column.

5.2.6 Bradford Protein Assay

Protein was quantified with the Bradford assay using bovine gamma globulin as the standard [31].

5.2.7 Electrophoresis

Sodium dodecyl sulfate polyacrylamide gel electrophoresis (SDS-PAGE) was performed as described by Laemmli [32]. The gels were BioRad mini-protean TGX precast gels 4–20 %. They were stained with Coomassie Blue.

5.2.8 Solid-Phase Extraction

A C-18 solid-phase extraction (SPE) cartridge (Alltech) was used to purify Fraction 2. One mg of Fraction 2 dissolved in 1 mL of water was applied to the column. The brown color was adsorbed to the cartridge and was eluted with 0.5 mL methanol.

5.2.9 FT-IR

The Fourier transform infrared (FT-IR) spectrum was obtained by scanning 16 times on a neat film and with a resolution of 16 cm^{-1} using a Shimadzu FT-IR Miracle ATR in reflectance mode.

5.2.10 Statistics

Differences between samples were tested for significance with one-way ANOVA.

5.3 Results and Discussion

5.3.1 Degradation of BPA by Little Bluestem Seeds

Little bluestem seeds were incubated in a solution containing BPA (46 mg/L) for 6 days. After 6 days, the amount of BPA remaining in the solution was determined and compared with the control solution that did not contain seeds. When seeds were present the amount of BPA in the solution was decreased by 65 % (Figure 5.1). To determine whether BPA was taken up by the seeds, the seeds were extracted, and the amount of BPA extracted from the seeds was 3 % of the control. This indicates that 62 % of the BPA was either metabolized or was taken up by the seeds and could not be extracted. There is precedence for BPA becoming inextractable following absorption as shown by a study that traced BPA's absorption and metabolism in a plant, water convolvulus (*Ipomoea aquatica*) [24]. The authors concluded that over 50 % of the absorbed BPA was either tightly bound to plant material or had polymerized. This is unlikely to be the case for little bluestem seeds, however, because the exudate from

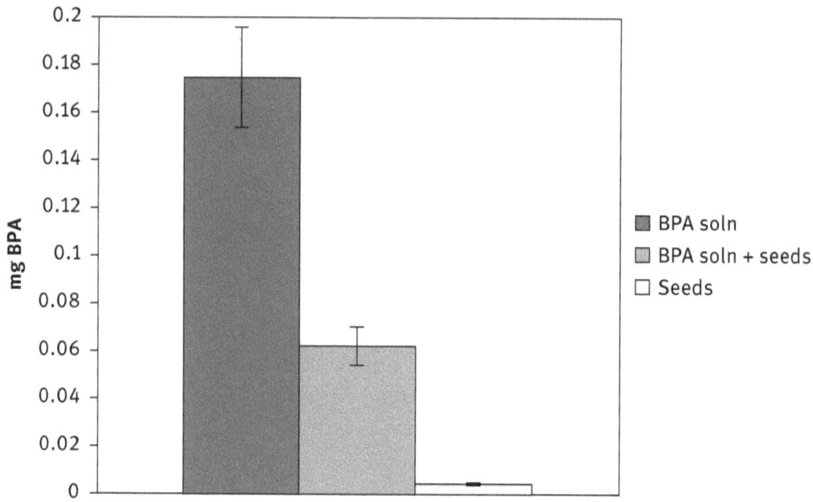

Figure 5.1: BPA solution (46 mg/L) was incubated with and without little bluestem seeds for 6 days, after which the solution was analyzed for the amount of BPA present. Seeds were extracted with methanol and the amount of BPA in the methanol extract was measured. The solid bar is the control (BPA solution without any seeds present), the hashed bar is the BPA solution that contained seeds, and the unfilled bar is the BPA extracted from seeds. Error bars represent the standard deviation of four replicates.

little bluestem seeds is able to degrade the BPA, and this would probably occur before the BPA could be incorporated into the seeds.

Of the BPA that was not recovered from seeds or solution (62 %), some of it may have been taken up by the seeds but was not extractable, some of it may have been taken up by the seeds and metabolized, and some of it may have been metabolized by something that leached out of the seeds, a seed exudate. The next experiment tested for the latter: can seed exudate degrade BPA? Seeds were soaked in water for 9 days and the aqueous solution, which now had a brown color, was collected and used as the seed exudate. Seed exudate was incubated with BPA to determine if it could metabolize BPA. The exudate caused a 74 % decrease in the amount of BPA after 5 days (Figure 5.2). To determine whether the active component in seed exudate was sensitive to heat, a portion of the exudate was boiled before it was incubated with BPA. Although there was some decrease in the amount of BPA, it was not significantly different than the control. This experiment indicates that little bluestem seeds exude something into the water that is capable of breaking down BPA. The active component in the exudate can be inactivated by heat. This is consistent with the idea that an enzyme is exuded from the seeds as they are germinating. A Bradford protein assay indicated that the exudate was 25 % protein by weight also supporting the hypothesis that it is an enzyme. Loffredo et al. [29] had similar results with seed exudate from

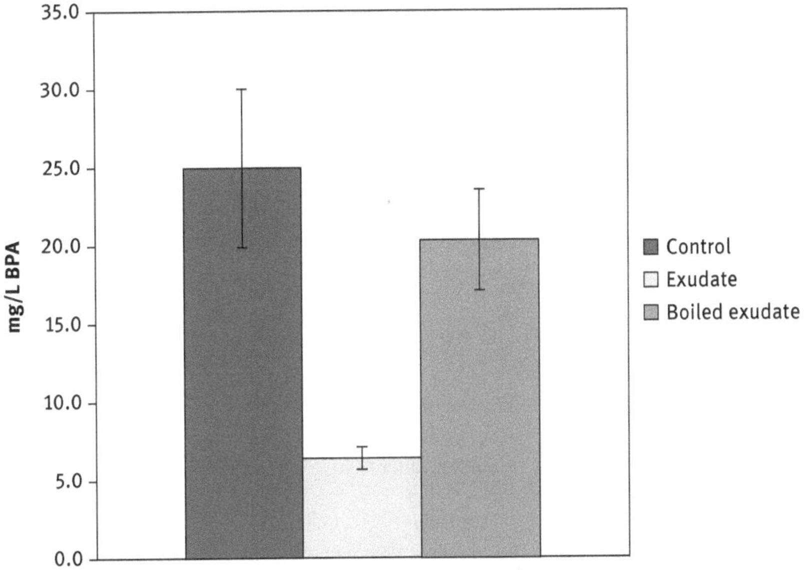

Figure 5.2: Little bluestem seed exudate was collected after 9 days. A portion of it was placed in boiling water for 10 min to give "boiled exudate." Exudate and boiled exudate were incubated with BPA (30 mg/L) for 5 days after which the solution was analyzed to determine the BPA concentration. The control contained water instead of exudate. Error bars represent the standard deviation of four replicates. The solid bar is the control, the unfilled bar is the BPA solution incubated with exudate, and the bar with diagonal lines is the BPA solution incubated with boiled exudate.

perennial ryegrass which was found to degrade BPA. They also hypothesized that it was enzymatic activity.

With the understanding that something in the exudate was able to degrade BPA, attempts were made to purify the substance. It was successfully separated by size exclusion chromatography (Sephadex G-75) to give two fractions (Figure 5.3). The second fraction was notable in that it was a very broad peak and was brown in color. The first fraction was clear. When standards were run on this column (blue dextran and DNP-aspartate), the total volume of the column was determined to be 28 mL. This means that even the smallest protein should elute in 28 mL. The elution volume of the second fraction from seed exudate was about 35 mL, suggesting that the compound was actually binding to the Sephadex G-75 rather than simply traveling through the pores of the size exclusion resin as would be expected for a small protein.

To determine which of these fractions contained the active component for degrading BPA, each fraction (1 and 2) was incubated with BPA to determine whether it could degrade BPA. Neither of the fractions alone could degrade BPA, but when they were combined, BPA degradation occurred (Figure 5.4). This indicates that both fractions must be present in order to degrade BPA.

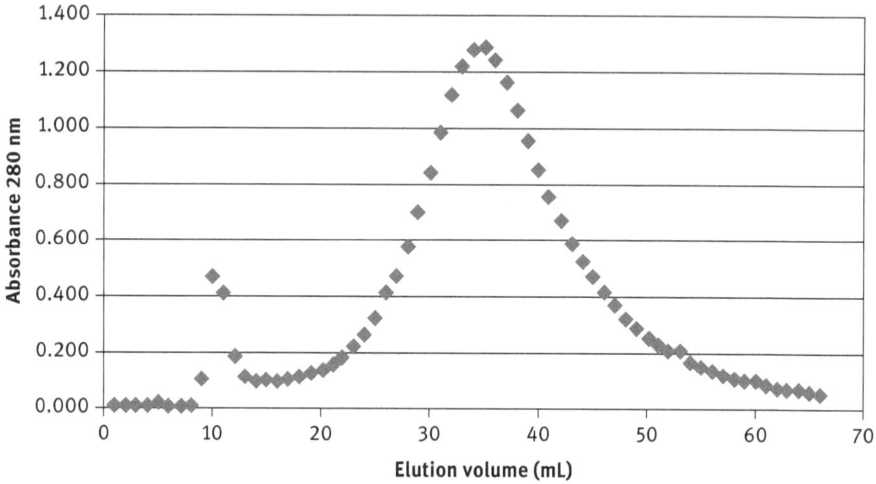

Figure 5.3: Little bluestem seed exudate (collected after 8 days and freeze-dried) was applied (3.7 mg) to a Sephadex G-75 column (34 × 2 cm). The column was eluted with 0.100 M potassium phosphate buffer pH 6.5. Eluant was collected in 1 mL increments and the absorbance was measured at 280 nm. The first eluting component is fraction 1 and the second one is fraction 2.

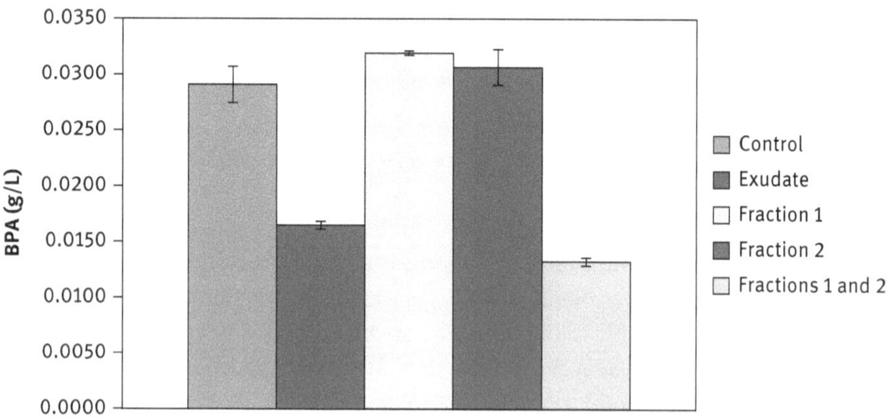

Figure 5.4: Little bluestem seed exudate (collected after 7 days) was freeze-dried, and a portion of it was separated by Sephadex G-75. Freeze-dried exudate and column fractions were incubated with BPA (30 mg/L). The concentration of BPA was determined after 6 days. The control contained water instead of exudate. Error bars represent the standard deviation of four replicates. The solid bar is the control, unfilled bar is freeze-dried exudate, bars with vertical and horizontal lines are fractions 1 and 2, respectively, and the bar with diagonal lines is a combination of the two fractions.

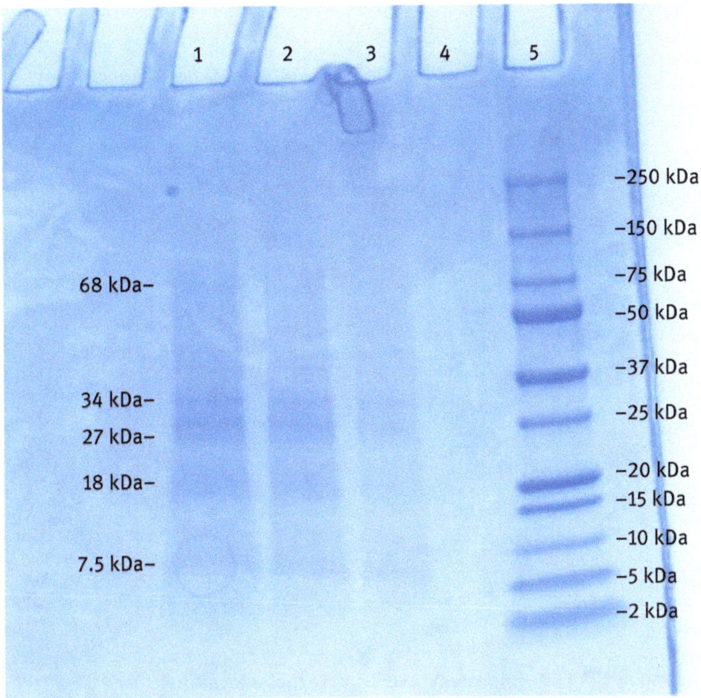

Figure 5.5: SDS-PAGE of fraction 1 from little bluestem seed exudate. Lanes 1, 2, and 3 are decreasing amounts of fraction 1, lane 4 is empty, and lane 5 is the molecular weight standards.

In order to characterize the components of fractions 1 and 2, they were subjected to electrophoresis. Fraction 1 was successfully separated by SDS-PAGE showing multiple protein bands with molecular weights ranging from 7.5 to 68 kDa (Figure 5.5). Fraction 2 could not be separated by SDS-PAGE because it appeared to interact with the gel in a way that inhibited its migration.

Fraction 2 is a brown substance that is very water soluble. Often it is polyphenols in plant extracts that impart a brown color. In this case, however, we know this compound is not phenolic because it was not positive with the Prussian Blue reagent [33]. To attempt to identify the substance in fraction 2, it was further separated using a solid phase C-18 extraction cartridge. Fraction 2 was applied to the cartridge, and then the brown color was eluted with methanol. The methanol eluent was analyzed with FT-IR (Figure 5.6). The IR spectrum indicates that fraction 2 does not contain protein because there is no carbonyl stretch (1,870–1,540 cm^{-1}). However, there is evidence that it contains an aromatic ring with C–H stretching around 3,100 cm^{-1}, C–H bending around 1,000 cm^{-1} and C=C ring stretching in the 1,500–1,400 cm^{-1} region. There is also evidence that it contains an amine with an N–H stretch at 3,402 cm^{-1}. The aromatic ring is also consistent with the compound's absorbance at 280 nm when it eluted from the Sephadex G-75 column.

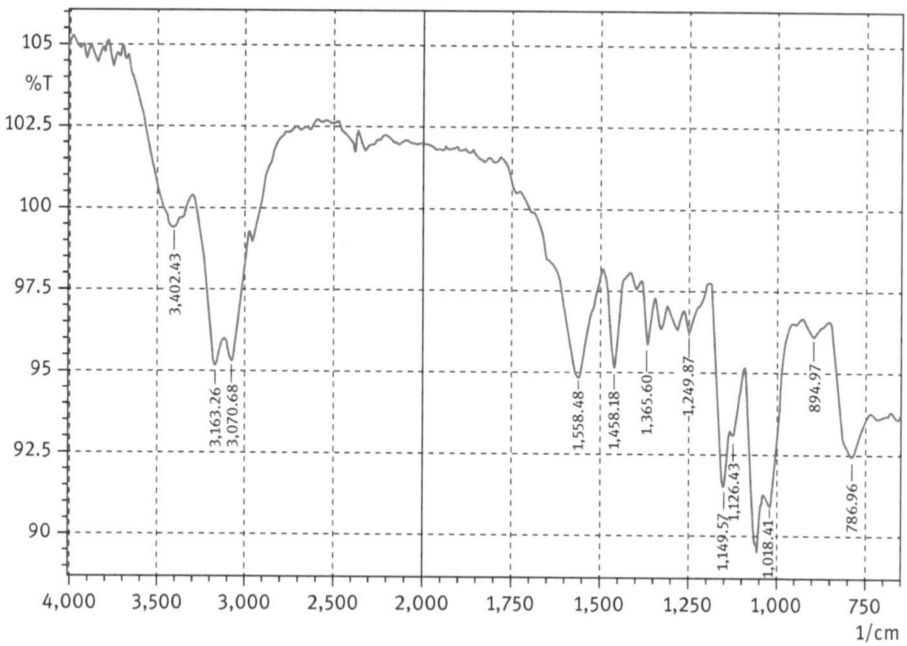

Figure 5.6: IR spectrum of fraction 2 of little bluestem seed exudate that was purified using a C-18 SPE cartridge.

It is likely that fraction 2 is some sort of cofactor needed by the protein in fraction 1 for the enzyme activity that degrades BPA. It remains a challenge to determine the structure of the cofactor and the source of the brown color.

5.4 Conclusion

In summary, little bluestem seeds can remove BPA from an aqueous solution as they are germinating. Most of the BPA is degraded by seed exudate that is present in the aqueous solution after the seeds have soaked in water for 6–10 days. The component of seed exudate that degrades BPA appears to be an enzyme that requires a cofactor that has yet to be fully characterized.

Little bluestem seeds have the potential to remove BPA from industrial waste effluent. They would need to be tested on a larger scale and in the presence of the other components of the effluent. The concept of using seeds to clean up contaminated water is also being pursued by Gattullo et al. [34] who used radish seedlings and showed that they could degrade BPA and other endocrine disrupters in samples of water. Alternatively, the enzyme from seed exudate could be isolated and used to degrade BPA in the effluent. Using enzymes in water treatment has been reported by Cabana et al. [35] who used a mixture of fungal enzymes to remove BPA from wastewater and Tsutsumi et al. [36] who used the enzyme laccase to successfully remove BPA and nonylphenol from water. Little bluestem seeds are readily available and it would be worth pursuing their use in cleaning up BPA from effluent.

References

[1] U.S. Environmental Protection Agency (USEPA). Bisphenol A Action Plan. 2010, https://www.epa.gov/sites/production/files/2015-09/documents/bpa_action_plan.pdf (accessed June 15, 2016).

[2] Merchant Research & Consulting Ltd. Bisphenol A (BPA): 2014 World Market Outlook and Forecast up to 2018. 2014, https://marketpublishers.com/report/industry/chemicals_petrochemicals/bisphenol_a_world_market_outlook_n_forecast.html (accessed June 10, 2016).

[3] Chemical Industry Education Center. Polycarbonates. *The Essential Chemical Industry Online*. 2013, http://www.essentialchemicalindustry.org/polymers/polycarbonates.html (accessed June 8, 2016).

[4] Eladak, S.; Grisin, T.; Moison, D.; Guerquin, M.-J.; N'Tumba-Byn, T.; Pozzi-Gaudin, S.; Benachi, A.; Livera, G.; Rouiller-Fabre, V.; Habert, R. A New Chapter in the Bisphenol A Story: Bisphenol S and Bisphenol F are Not Safe Alternatives to This Compound. *Fertil Steril* 2015, *103* (1), 11–21.

[5] Rochester, J. R.; Bolden, A. L. Bisphenol S and F: A Systematic Review and Comparison of the Hormonal Activity of Bisphenol A Substitutes. *Environ Health Perspect* 2015, *123* (7), 643–650.

[6] Reno, K. H.; Stanzione, J. F. III; Epps, T. H. III; Wool, R. P. Tackling Toxicity: Designing a BPA Alternative from Lignin. *Abstracts of Papers, 247th ACS National Meeting & Exposition, Dallas, TX, United States*, March 16–20, 2014.

[7] Epoxy Resin Committee. Epoxy Resins: Assessment of Potential BPA Emissions-Summary Paper. 2015, http://www.epoxy-europe.eu/uploads/Modules/Resources/epoxy_erc_bpa_whitepapers_summarypaper.pdf (accessed June 8, 2016).

[8] Makinwa T. T.; Uadia, P. O. A Survey of the Level of Bisphenol A (BPA) in Effluents, Soil Leachates, Food Samples, Drinking Water and Consumer Products in South-Western Nigeria. *World Environ* 2015, *5* (4), 135–139.

[9] Cousins, I.T.; Staples, C. A.; Klečka, G.M.; Mackay, D. A Multimedia Assessment of the Environmental Fate of Bisphenol A. *Hum Ecol Risk Assess* 2002, *8*, 1107–1135.

[10] Dorn, P. B.; Chou, C.S.; Gentempo, J. J. Degradation of Bisphenol A in Natural Waters. *Chemosphere* 1987, *16* (7), 1501–1507.

[11] Klecka, G. M.; Gonsior, S. J.; West, R. J.; Goodwin, P. A.; Markham, D. A Biodegradation of Bisphenol A in Aquatic Environments: River Die-Away. *Environ Toxicol Chem* 2001, *20* (12), 2725–2735.

[12] Kolpin, D. W.; Furlong, E. T.; Meyer, M. T.; Thurman, E. M.; Zaugg, S. D.; Barber, L. B.; Buxton, H. T. Pharmaceuticals, Hormones, and Other Organic Wastewater Contaminants in US Streams, 1999–2000: A National Reconnaissance. *Environ Sci Technol* 2002, *36* (6), 1202–1211.

[13] Marcial, H. S.; Hagiwara, A.; Snell, T. W. Estrogenic Compounds Affect Development of Harpacticoid Copepod *Tigriopus japonicus*. *Environ Toxicol Chem* 2003, *22* (12), 3025–3030.

[14] Kashiwagi, A.; Utsumi, K.; Kashiwagi, K.; Ohta, S.; Sugihara, K.; Hanada, H.; Kitamura, S. Effects of Endocrine Disrupting Chemicals on Amphibian Metamorphosis and Mitochondrial Membrane Permeability Transition. *J Health Sci* 2008, *54* (3), 273–280.

[15] Flint, S.; Markle, T.; Thompson, S.; Wallace, E. Bisphenol A Exposure, Effects, and Policy: A Wildlife Perspective. *J Environ Manag* 2012, *104*, 19–34.

[16] Environment Canada, Consultation Document Phenol, 4,4' -(1-Methylethylidene)bis-(Bisphenol A). 2009, http://www.ec.gc.ca/lcpe-cepa/79A93C66-1EE0-4F72-89C1-114DFD7B7292/pba-eng.pdf (accessed June 13, 2016).
[17] Canesi, L.; Fabbri, E. Environmental Effects of BPA: Focus on Aquatic Species. *Dose-Response* 2015, *13* (3), 1–14.
[18] Horikoshi, S.; Tokunaga, A.; Hidaka, H.; Serpone, N. Environmental Remediation by an Integrated Microwave/UV Illumination Method: VII. Thermal/Non-thermal Effects in the Microwave-Assisted Photocatalyzed Mineralization of Bisphenol-A. *J Photochem Photobiol A: Chem* 2004, *162* (1), 33–40.
[19] Cabana, H.; Jones, J. P.; Agathos, S. N. Elimination of Endocrine Disrupting Chemicals Using White Rot Fungi and Their Lignin Modifying Enzymes: A Review. *Eng Life Sci* 2007, *7* (5), 429–456.
[20] Lobos, J. H.; Leib, T. K.; Su, T. M. Biodegradation of Bisphenol A and Other Bisphenols by a Gram-Negative Aerobic Bacterium. *Appl Environ Microbiol* 1992, *58* (6), 1823–1831.
[21] Hamada, H.; Tomi, R.; Asada, Y.; Furuya, T. Phytoremediation of Bisphenol A by Cultured Suspension Cells of *Eucalyptus perriniana* – Regioselective Hydroxylation and Glycosylation. *Tetrahedron Lett* 2002, *43*, 4087–4089.
[22] Nakajima, N.; Ohshima, Y.; Serizawa, S.; Kouda, T.; Edmonds, J.; Shiraishi, F.; Aono, M.; Kubo, A.; Tamaoki, M.; Saji, H.; Morito, M. Processing Bisphenol A by Plant Tissues: Glucosylation by Cultured BY-2 Cells and Glucosylation/ Translocation by Plants of *Nicotiana tabacum*. *Plant Cell Physiol* 2002, *43* (9), 1036–1042.
[23] Imai, S.; Shiraishi, A.; Gamo, K.; Watanabe, I.; Okuhata, H.; Miyasaka, H.; Ikeda, K.; Bamba, T.; Hirata, K. Removal of Phenolic Endocrine Disruptors by *Portulaca oleracea*. *J Biosci Bioeng* 2007, *103* (5), 420–426.
[24] Noureddin, I. M.; Furumoto, T.; Ishida, Y.; Fukui, H. Absorption and Metabolism of Bisphenol A, a Possible Endocrine Disruptor, in the Aquatic Edible Plant, Water Convolvulus (*Ipomoea aquatica*). *Biosci Biotechnol Biochem* 2004, *68* (6), 1398–1402.
[25] Saiyood, S.; Vangnai, A.; Thiravetyan, P.; Inthorn, D. Bisphenol A Removal by Dracaena Plant and the Role of Plant-Associating Bacteria. *J Hazard Mater* 2010, *178*, 777–785.
[26] Tober, D.; Jensen, N. Plant Guide for Little Bluestem (*Schizachyrium scoparium*]. *USDA Natural Resources Conservation Service, Plant Materials Center, Bismarck, ND*, 2013.
[27] Pradhan, S. P.; Conrad, J. R.; Paterek, J. R.; Srivastava, V. J. Potential of Phytoremediation for Treatment of PAHs in Soil at MGP Sites. *J Soil Contam* 1998, *7* (4), 467–480.
[28] Frick, C. M.; Farrell, R. E.; Germida, J. J. Assessment of Phytoremediation as an In-situ Technique for Cleaning Oil-Contaminated Sites. Submitted to *Petroleum Technology Alliance of Canada (PTAC)* 1999, https://rtdf.clu-in.org/public/phyto/phyassess.pdf (accessed June 13, 2016).
[29] Loffredo, E.; Gattullo, C. E.; Traversa, A.; Senesi, N. Potential of Various Herbaceous Species to Remove the Endocrine Disruptor Bisphenol A from Aqueous Media. *Chemosphere* 2010, *81* (1), 1274–1277.
[30] Gatidou, G.; Thomaidis, N. S.; Stasinakis, A. S.; Lekkas, T. D. Simultaneous Determination of the Endocrine Disrupting Compounds Nonylphenol, Nonylphenol Ethoxylates, Triclosan and Bisphenol A in Wastewater and Sewage Sludge by Gas Chromatography–Mass Spectrometry. *J Chromatogr A* 2007, *1138* (1–2), 32–41.
[31] Bradford, M. M. A Rapid and Sensitive Method for the Quantitation of Microgram Quantities of Protein Utilizing the Principle of Protein-Dye Binding. *Anal Biochem* 1976, *72* (1/2), 248–254.
[32] Laemmli, U. K. Cleavage of Structural Proteins During the Assembly of the Head of Bacteriophage T4. *Nature* 1970, *227* (5259), 680–685.
[33] Roland, B.; Tonelli, D.; Girotti, S. Analysis of Total Phenols Using the Prussian Blue Method. *J Agric Food Chem* 1980, *28*, 1236–1238.

[34] Gattullo, C. E.; Cunha, B. B.; Loffredo, E.; Rosa, A. H.; Senesi, N. Phytoremediation of the Endocrine Disruptors Bisphenol A, Linuron and 17α-Ethinyl estradiol in NOM-Enriched Water and Freshwaters, in *Functions of Natural Organic Matter in Changing Environment*, Xu, J.; Wu, J.; He, Y. (eds.). Springer Netherlands: Dordrecht, 2013, 607–610.

[35] Cabana, H.; Jones, J. P.; Agathos, S. N. Elimination of Endocrine Disrupting Chemicals using White Rot Fungi and their Lignin Modifying Enzymes: A Review. *Eng Life Sci* 2007, *7* (5), 429–456.

[36] Tsutsumi, Y.; Haneda, T.; Nishida, T. Removal of Estrogenic Activities of Bisphenol A and Nonylphenol by Oxidative Enzymes from Lignin-Degrading Basidiomycetes. *Chemosphere* 2001, *42* (3), 271–276.

Larry Kolopajlo

6 Green Chemistry Pedagogy

Abstract: This chapter attempts to show how the practice of chemistry teaching and learning is enriched by the incorporation of green chemistry (GC) into lectures and labs. To support this viewpoint, evidence from a wide range of published papers serve as a cogent argument that GC attracts and engages both science and nonscience students, enhances chemistry content knowledge, and improves the image of the field, while preparing the world for a sustainable future. Published pedagogy associated with green and sustainable chemistry is critically reviewed and discussed.

Keywords: green chemistry education, green chemistry pedagogy

6.1 Introduction

If chemistry is the central science, then green chemistry (GC) is central to sustainability. What's more, although sustainability is a global approach to solving the world's present and future problems, GC, now sometimes called sustainable chemistry, being a localized strategy, alone cannot achieve the goal, but must be one of a host of scientific, economic, social, and political approaches. A transformation to a sustainable world not only requires highly skilled chemists, but additional professionals from a broad range of interdisciplinary fields. Hence, versatile and qualified instructors must be available and equipped to teach GC and sustainability literacy to a wide audience of students from elementary through graduate school. But how can teachers be recruited, effectively trained, and equipped with the right educational tools to match the need for green chemical education (GCE)? To answer this question, one first examines the status of science employment and education.

Studies [1, 2] have indicated that a strong technical and professional science, technology, engineering, and mathematics (STEM) workforce, trained through high-quality K-12 and postsecondary education, will significantly impact the nation's competitiveness and economic stability. Moreover, STEM jobs are generally well paid, and often have low unemployment rates, being about half that of non-STEM occupations. For example, according to Rothwell [3], about 20 % of all US jobs, equivalent to 26 million, require significant STEM skills.

But the United States is not graduating enough of its own citizens to fill jobs in chemistry and there is a dire need to train more scientists, including chemists. To fill the gap, US graduate schools recruit international students, who, according to a 2008 demographics report [4] issued by the American Chemical Society (ACS), make up 42 % of the graduate chemistry population. Many jobs are open because the unemployment [5] among chemists currently hovers around 3 %, much lower than in the general population. According to the U.S. Department of Education Report [6] titled Education for Global Leadership, STEM jobs are projected to increase by 14 % between 2010 and 2020, but fewer than 20 % of all high school students are interested in STEM.

Studies indicate that too many US students are neither prepared for nor interested in STEM. One standardized measure of US high school student science achievement is the Program for International Student Assessment or PISA [7], which measures achievement in reading literacy, mathematics, and science literacy for 15-year-olds in about 65 countries across the globe. Overall, the 2012 PISA results in science literacy [8] show that the United States is in 23rd place, and its average score of 497 is neither measurably different from the Organisation for Economic Co-operation and Development average of 501, nor measurably different from 13 competing educational systems.

The science literacy conundrum facing the United States is: Why is high school interest and PISA student performance stagnant when there is such a need for highly skilled, highly paid, and innovative scientists? For chemistry, one problem might be the negative images attributed to it due to environmental disasters like Love Canal [9]. More recently, the fact that many companies have adopted green practices such as reformulating products in response to green consumer demand, for economic benefits, and because of their commitment to social welfare and the environment, implies that chemistry and science college students would be better prepared for an industrial role if they received some GC training. Therefore, GC may provide a way to not only improve the public image of chemistry, but to engage students, promote K-12 science literacy, and attract them to the chemistry profession. Integrating GC into the K-12 and college teaching requires that K-12 educators and science education professionals at both the undergraduate and graduate levels be better versed in the subject. These educators then must determine how GC can fit into the K-12 classroom. Hence, it is worthwhile to investigate pedagogy associated with GC.

In conventional terms, pedagogy [10] refers to the art and science of teaching, encompassing the best practices regarding both content and methods. Although pedagogy refers to both teacher- and student-centered activities, it more often refers to a teacher's role in a classroom or laboratory regarding style, context, and the selection of what instructional materials and approaches are appropriate and effective, taking into account prior knowledge of students and goals of learning. It involves what, when, how, and why something is taught. However, pedagogy is more than planning, setting learning outcomes and objectives, and other technical factors regarding the arrangement, and execution of classroom activities. In a more modern or contemporary sense, it serves as a deployment model for classroom teaching in which teachers are viewed as facilitators of learning.

Current research, taking into account the science of how people learn, has shown that classroom pedagogy should shift from an expository model of knowledge transmission to inquiry and constructivist instruction [11]. George Bodner [12, 13] summed up constructivism by writing, "Knowledge is constructed in the mind of the learner."

One constructivist pedagogy often recommended is inquiry, which can have many forms, from guided to open [14–16].

Inquiry pedagogy involves all of the decisions that teachers make, regarding content and methods, to promote science instruction to train students to use the scientific method, in the way it is practiced by scientists in research. Therefore, inquiry involves all of the science process and content skills associated with the scientific method. There are three problems that must be overcome to make science instruction effective. First, students must be attracted to the profession of science; second, they must be engaged in classrooms; and finally they must understand and learn science content. To engage students, in his book, *Creating Significant Learning Experiences*, L. Dee Fink [17] describes how to transform a traditional classroom into one that engages students, and increase student interaction and satisfaction through the human dimension of learning. Fink's Significant Learning Model takes into account the affective side of learning, and it would fit well with teaching strategies to disseminate the cognitive side of GC.

The field of GC is perhaps the best example in chemistry of how to make chemistry relevant to a wide audience, while improving its image. Moreover, in association with GC content, the parallel field of GC pedagogy, referring to both the teaching and learning strategies based on the principles of GC, that occurs in the classroom or laboratory is important as well. But GC is an emerging field whose teaching methodologies have not been well documented because they have only evolved over the last 20 years. Because GC is expanding, and many potential instructors need training, successful examples of instructional strategies or pedagogy associated with GC are worth reviewing.

6.2 GC Reviews

Only a few reviews on green chemistry education (GCE) exist, and so there is a need to elaborate on them. Levy et al. [18] reviewed much of the published GCE work through 2005, and although they insisted that the GC community was developing an extensive array of cross-curricula resources, few studies outside of the organic domain have been published by 2015.

Andraos and Dicks [19] published a critical and comprehensive review of GC-effective educational practices, including teaching resources both online and in print, emphasizing green organic chemistry. This review mainly addressed work at the undergraduate level through 2011. According to the authors, there are GC pedagogical strengths and weaknesses in GC instruction, as noted in Table 6.1 where special attention is paid to organic GC.

Although the work of Andraos and Dicks is the most valuable GCE review to date, the field is expanding and evolving at such a fast pace that their work requires elaboration. For example, no review has addressed GCE pedagogy at the high school

Table 6.1: Strengths and weaknesses of GC education according to Andraos and Dicks.

Pedagogical benefits of GCE	Pedagogical weaknesses in GCE
1. GC can be either integrated with existing courses or be stand-alone	1. More work must be published in subfields beyond organic chemistry
2. Many real-world case studies exist, like the Presidential Green Chemistry Award Challenges, to make GC relevant	2. Metrics published in the literature are neither unified nor standardized
3. GC is both a qualitative and quantitative field that allows green decision-making	3. Good textbooks are lacking, especially in the organic subfield
4. There are many lab resources available for organic chemistry	4. Few courses exist and are usually electives
5. Many excellent web-based materials, case studies, and journal articles provide valuable resources	5. Not all published work meets green criteria, including some that are catalytic, single pot, solventless, or that use ionic liquids
6. Lab courses remain an excellent vehicle for teaching GC principles	6. Published papers often lack enough experimental detail to enable exact replication of syntheses
GC benefits for undergraduates Students: 1. achieve a deeper and critical understanding of organic synthesis	7. Published claims are not always supported by data regarding metrics, life cycle assessments, and energy consumption
2. are better trained and prepared to work in industry	8. In organic reactions, more examples of new protecting groups, catalyst recycling, atom economy in redox reactions, and the use of water as a reaction solvent are needed
3. retain more content knowledge and understand it better because of higher motivation	

level in methods, content, or outreach. Moreover, little work addresses the pedagogy of undergraduate GC courses in the liberal arts and general chemistry. Last of all, no review has addressed recent papers that seem to shift from using "green chemistry" to "sustainable chemistry," and describe new socioscientific approaches to teaching GC together with sustainability.

This review of GCE consists of two parts. Part 1 covers much of what has not been addressed in previous reviews, especially elaborating on GC in general chemistry and liberal arts labs and courses, emphasizing pedagogy by summarizing studies published in the *Journal of Chemical Education* [20], *Chemical Education Research and Practice* (*CERP*) [21], and *The Chemical Educator,* and journals issued by the National Science Teachers Association (NSTA) [22]. Since Andraos and Dicks published a thorough review of organic GC, only innovative pedagogies for organic that may be transferred to other areas will be covered. The emerging subfields of GC nanochemistry and GC analytical will not be covered. This review begins by addressing high school GC, and then proceeds through general chemistry GC and organic pedagogy, and finally covers sustainable chemistry education in part 2.

6.3 Part 1: GC Courses and Lab Pedagogy

6.3.1 GC Academic Programs

Over the last 20 years there has been an explosive growth in GC courses and programs. The article titled "The Ivory Tower Goes Green" [23] describes how academic programs infused GC into chemistry curricula in 2008. Fast-forwarding to 2015, according to the ACS GC Web site [24], there are now 41 GC academic programs in 25 US states. However, this is a fledgling number since according to the ACS [25], there are 681 ACS-approved chemistry programs in the United States. Internationally, GC also surged academically, and by 2015 there were more than 33 GC programs offered by 16 countries and organizations.

6.3.2 High School GC

Very little has been published regarding GC in the secondary curriculum, and there is a need for novel activities, experiments, and case studies. One way to publicize GC in high school is through outreach. GC outreach in the K-12 system is becoming more common as described in a write-up on *Beyond Benign* [26], a foundation that advances GC in New England. National Chemistry Week "Celebrating Chemistry" pamphlets [27] have touched on GC in its outreach for Mole Day. GC outreach is also supported by the Green Chemistry Institute [28].

In high school chemistry curricula, although the term "green chemistry" has yet to be written into NGSS standards [29], at least "sustainability" appears. The NSTA has published a few quality GC papers in its journals that cover both K-12 science teaching and college teaching. Most of the NSTA articles regarding GC center on stimulating student interest in chemistry while imparting a positive image of the field and its relevance to everyday life. For example, in its "Career of the Month" column, Megan Sullivan [30] describes what its like to be a Green Product Chemist, by summarizing her interviews with several green chemists. She provides a very relevant and inspiring example in delineating how Nike Corporation used professionally trained R&D green chemists to develop a "greener" rubber by identifying and replacing hazardous chemicals with naturally sourced chemicals.

Ken Roy also authored a number of articles touching on GC. In the column titled "Scope on Safety," in the article, "Greener Is Cleaner and Safer," Ken Roy [31] informs readers of the 12 principles of GC, in an effort to reduce risks and accidents, especially in middle school. In another article titled "Safer Science," Roy [32] discusses sustainable safety practices being encouraged by the EPA (Environmental Protection Agency) that cover GC procedures on how to properly select and use commercial cleaning products with the intent of decreasing fumes and particulates that cause indoor air pollution. In another article titled, "Is Green Cleaner," Roy [33] argues that "students at the middle school level are at greater risk when exposed to toxic chemicals."

A 2010 paper titled "Green Root Beer Laboratory [34]" appears to be the first actual GC lab published in an NSTA journal. Students formulate a green root beer using local products (root beer extract) and recycled materials (plastic bottles). Content knowledge addressed includes using acid–base chemistry and two brewing methods: (a) dry ice and (b) yeast. Unfortunately, the latter process produced up to 0.5 % ethanol rendering the product unfit for child consumption.

However, in an NSTA Green Science column article [35] "Green Beauty," the author writes that "many cosmetics contain ingredients that are linked to health problems and environmental concerns" and issues one warning after another related to chemicals that the Food and Drug Administration (FDA) deems safe. The author's opinion that oxybenzone in sunscreens is dangerous is overstated because not only does the FDA consider it safe, but more importantly, it reduces skin cancer risk caused by the sun's UV radiation. When teaching the same material, a green chemist might instead invoke a risk/reward strategy in evaluating the value of oxybenzone in sunscreens. Using "green" to misinform a young audience may induce chemophobia at an early age.

Although Mandler et al. [36] did not explicitly focus on GC, their work arguably promotes both green and sustainable literacies by devising and deploying a unit that integrated environmental and analytical chemistry, thus transforming content into a relevant and real-world context. The unit "I Have Chemistry with the Environment" consisted of two modules, one on drinking water quality, and the second on the greenhouse effect. Results indicated that students underwent positive changes in both attitudes toward chemistry and were better motivated to learn chemistry.

6.3.3 College General Chemistry

Little has been written regarding how to incorporate GC into undergraduate general chemistry and therefore this section will review in depth GC pedagogy used in labs, lecture, a case study, and a demonstration. Generally speaking, published work involves two kinds of pedagogy: (a) the chemistry content subfield pedagogy which is primary and (b) the GC content pedagogy which is ancillary, but which also carries the function of motivation. Hence, GC crosses the cognitive mode and also enhances the affective side by increasing motivation. In short, GC primarily serves as a vehicle to teach chemistry subfield content, and secondarily GC content.

6.3.3.1 Example 1: Lab

Cacciatore and Sevian [37] published a green-centered experiment to teach stoichiometry to AP (Advanced placement) or general chemistry college students. Besides teaching three GC principles related to atom economy, safer chemicals, and waste prevention, students used inquiry to discover stoichiometry concepts while learning why and how to write a laboratory report. Regarding pedagogy, this paper also provides a clear distinction between cookbook and inquiry labs. In a prelab exercise, students performed a stoichiometry calculation similar to the one they would perform in lab. The prelab exercise addressed the principles of GC, student prior knowledge concerning stoichiometry, allowed students time to correct misconceptions, and provided a scaffold to support the actual lab work to follow. The actual experiment involved determination of the composition of a mixture of sodium carbonate and sodium bicarbonate by heating and weighing. Students worked in groups of two, and each group was provided one of three different sample lab reports, each report having one of the following deficiencies: (a) no materials or procedure, (b) nonreproducible results, and (c) no discussion section. Students had to confirm or refute the data given in their sample reports by performing the experiment. Student groups were permitted to discuss the experiment with other groups but could not exchange lab reports. While doing the lab, students practiced inquiry, critical thinking, the scientific method, constructed new knowledge, and avoided applying a memorized algorithmic approach to stoichiometry. GC principles were applied in the following ways: (a) all waste was recycled and reused, (b) neither sodium carbonate nor sodium bicarbonate is hazardous, and (c) all of the reactants converted to products illustrating atom economy. Students responded positively to the lab indicating the power of GC as a tool to stimulate learning in high school or college general chemistry.

6.3.3.2 Example 2: Lab

Cacciatore et al. [38] reported a novel introductory GC laboratory experiment on solubility equilibrium and periodicity for an undergraduate general chemistry course. The experiment is also suitable for a high school advanced placement course. Although many solubility equilibrium experiments have been published for the general chemistry lab, this was the first to incorporate GC and periodicity. This experiment is also substantially greener than many previously reported solubility experiments because it does not employ hazardous heavy metals like barium, lead, and silver. The GC attributes pertaining to this experiment include using salts that are inexpensive, of low toxicity, and that pose little risk to students or to the environment. In addition, dilute solutions were used, so only a small amount of waste was produced; moreover, that waste was neutralized before disposal.

Students investigated solubility equilibria of sparingly soluble salts, including $Ca(OH)_2$, $Mg(OH)_2$, and $Sr(OH)_2$, and determined their K_{sp} values. The experimental procedure involved using a standardized HCl solution to titrate saturated solutions of the group(II) hydroxides to a phenolphthalein endpoint. From the collected data, students computed K_{sp} values for each salt. Since each salt contained only three ions, the K_{sp} computation did not lead to a cognitive overload. Next, students deduced periodic trends in solubility results, and applied and extended their constructed knowledge to predict the relative solubility of similar compounds. Finally, students demonstrated why their experiment fit in with GC principles.

The experimental design employed two different kinds of inquiry. The rarely used open inquiry design was used for the experimental part of the experiment, but a guided inquiry format was used for the lab report. Instead of using a cookbook procedure, students both planned and performed the experiment. But for the lab report, students completed partly written reports designed to guide students to use their solubility results to find a periodic trend and predict the solubilities of group(II) hydroxide salts not experimentally investigated.

To prepare students for the experiment, in a prelab exercise, students explored GC, solubility, solubility calculations, and periodicity. Next, when students came to lab, instead of starting work alone, they were arranged into groups and given one of three partially written lab reports. Although each lab report contained experimental data, each one also contained a different error or problem: (a) improper calculations, (b) incomplete procedure, or (c) no materials being listed. Besides correcting or filling in missing information, student groups had to confirm unverified experimental data through lab work. During the experiment, student groups were allowed to interact with other groups, but could not exchange or show their sample lab reports. This setup promoted collaboration and discussion.

The laboratory protocol utilized research, suggesting that science learning is promoted in environments that prompt students to construct their own knowledge through inquiry and discovery. This approach contrasts markedly with cookbook approaches where students follow set procedures written like recipes to verify a known experimental parameter. Before adopting the experiment for all general chemistry lab students, it was pretested with small groups of second year students, and with one pilot class of 30 general chemistry students. Both the prelab and experiment were judged a success. From a chemistry content perspective, almost all students performed the prelab computations correctly. Students also responded positively about the green attributes of the experiment and critically commented that many of their previous experiments were not green. In addition, student interaction and collaboration increased significantly compared to traditional setups. In the cognitive domain, students reported that compared to traditional experiments, this format entailed more thought and work. On the negative side, student evaluations showed that some students had trouble adjusting to the new format and to the partly written lab reports.

Other studies [39–41] have demonstrated that learning new concepts like solubility is promoted in situations where students integrate new with prior knowledge in a problem-solving environment. Moreover, new learning is also stimulated when students integrate different chemistry concepts like solubility, periodicity, and equilibrium.

6.3.3.3 Example 3: Lecture

Song et al. [42] developed a set of GC stoichiometric problems suitable for general chemistry lecture that were used to teach these principles of GC: atom economy, renewable feedstock as substitutes, safer chemistry for accident prevention, increased energy efficiency, and resource conservation. Classroom deployment involved giving students chemical reaction problems containing two or three competing reaction schemes, and asking them to find the optimized pathway by applying GC principle criteria. This instructional pedagogy utilized *guided inquiry* involving these *application of knowledge* questions:

1. Which reaction minimizes feedstock?
2. Which route minimizes pollution?
3. Which reaction reduces exposure to hazardous chemicals?
4. Which route optimizes the atom utilization ratio?
   and *synthesis* questions involving *higher order thinking* that draw on several GC principles to answer a question like:
5. Propose a new reaction that can save more chemicals.

Student solutions led to generalizations like: "when the feedstocks and reaction conditions are the same, a reaction using less feedstock to prepare a mole of products is better." In summary, by solving some green chemical reaction problems suitable for a college classroom, not only did students gain higher order thinking skills in GC and chemistry content, but they bridged the gap between education and industry.

6.3.4 GC in Other Papers

In another paper, Klingshirn et al. [43] published a green experiment to determine the formula of a metal hydrate, the authors choosing copper(II) chloride to replace the more dangerous barium salt.

In a second paper, a greener version [44] of the traditional Blue Bottle demo was developed. Optimizing the demonstration involved substituting sodium bicarbonate for potassium hydroxide, making the solution less caustic. In addition, ascorbic acid substituted for glucose, and additional copper sulfate pentahydrate was added. The modified version was not only safer, but it also reduced the amount of waste.

A green colligative property experiment [45] suitable for general chemistry class was reported in 2005. Traditional dangerous aromatic solvents like naphthalene are replaced by fatty acids like stearic acid, to measure the molar masses of solutes like lauric, myristic, and palmitic acids by freezing point depression. All of the fatty acids employed were nonhazardous, inexpensive, and had properties similar to their traditional counterparts. GC principles were demonstrated by using fatty acids originally derived from biomass rather than petroleum, and recovering and recycling them for further use so that waste was reduced or eliminated.

In summary, only a few papers have addressed GC high school and GC general chemistry, and there is a need for more work to be published. One possible avenue would be to integrate GC into the Nature of Science for science education students. GC case studies can also be built from published papers such as Jansen's [46] on the cost of converting a gasoline-powered vehicle to propane.

6.3.5 Courses and Curricula

In this section, we will be reviewing published work regarding GC courses and curricula. "Pre-curriculum" emergence of GC began in 1998 [47] when a collaboration between the ACS and the EPA created the Green Chemistry Educational Materials Development Project. The focus of many early papers such as the one published in 2000 by Hjeresen et al. [48] was to provide general information on GC, explain why it is important, and connect it to the environment. In some early papers covering GC chemistry curricula, some researchers advocated the view that GC was not a separate field of chemistry but a common component of all fields, more akin to ethical conduct, and a part of social and scientific responsibility. Another idea advanced regarding GC in the curriculum was that stand-alone GC courses were not viable as electives because they would draw few students, they were expensive startups, and they required trained faculty. Therefore, in most situations the best choice was to integrate GC within existing courses, which was the path chosen by Cann and Dickneider [49] who in 2004 published an article explaining how GC can be integrated into an existing curriculum. Their "infusion" method bypassed the rigor of new course scrutiny, and efficiently incorporated real-world GC examples into several courses in the chemical curriculum, including general and organic chemistry.

Later, in 2006, Braun et al. [50] published a commentary paper that originated from the July 2005 third annual ACS Green Chemistry Summer School (GCSS) held in Montréal, Québec, Canada. The paper also advanced the viewpoint that GC is not a distinct discipline or separate field of chemistry, but instead serves as a guiding principle for social and scientific responsibility leading to sustainability. According to the authors, as such, academic chemists should perform the difficult task of modifying their curricula so that GC becomes an important component. By doing so, they could strengthen the partnership between interdisciplinary scientific fields. Over the years as GC gained ground, GC courses became stand-alone and instances of GC programs became widespread. Obviously, GC has evolved from being considered ethical conduct into stand-alone programs across the globe.

6.3.6 GC Courses

Collins [51] describes the first GC course published in *Journal of Chemical Education* (*JCE*). The course titled "Introduction to Green Chemistry" demonstrated how GC evolved out of environmental programs at Carnegie Mellon University. The course was taught in 1992 and 1993 to upper-level chemistry undergraduates and graduate students as part of an environmental initiative across the university curriculum. The author noted the power of GC to serve as a vehicle imparting new relevance to chemistry. The course also helped students view chemistry as a positive force in a field whose name was often synonymous with pollution. Beyond stressing the 12 principles of GC, the course offered a critical analysis of the role of green reagents and catalysts. Moreover, a major segment of the course presented GC as an applied science, and therefore current problems facing humanity such as recycling of plastics and vulcanized rubber, replacing chlorofluorocarbon refrigerants, and Taxol synthesis were addressed. Students wrote and presented a technological proposal on one of the assigned GC topics.

In 1999, to instill an appreciation of the value of green industrial chemistry, Cann's [52] students made posters of award-winning Presidential Green Chemistry Challenge projects in an environmental chemistry course. The poster exhibition involved presentations and discussion.

In 2013, Manchanayakage [53] described a new GC course "Green Chemistry," offered as an upper-level elective for science majors at a liberal arts undergraduate institution. It was a two-credit course taken over a 14-week semester that coordinated both lecture and laboratory in a workshop format so that topics were coordinated across lecture and lab. One course goal was to allow an interdisciplinary science audience to investigate, collaborate, and discuss GC from different perspectives. It was hoped that this format would spur participants to implement greener solutions for the benefit of sustainability when they become professionals. The five main sections of the course organization that heavily emphasize green organic chemistry are shown in Table 6.2.

In the same paper, Manchanayakage described a new liberal arts GC course "Green Chemical Concepts" for nonmajors. The course was integrated into the liberal arts curriculum. It was a four-credit course taken over a 14-week semester that encompassed separate sections of lecture and laboratory. The course brought together students from diverse nonscience backgrounds and increased their awareness of how GC contributes to sustainability. It also trained students to be science literate and enabled them to make informed decisions on science policy and business. Being designed for small classes, the delivery method was described as "interactive lecture," employing

Table 6.2: Organization of GC course designed for multidisciplinary science majors.

1. Introduction to green chemistry
   Students study the 12 GC principles and apply them to the bromination of stilbene.
2. Environmental impact of chemical processes and products
   Students study waste management and LCA in the synthesis of substituted chalcones.
3. Alternative reaction media
   Students study solvent-free systems and perform a solventless Aldo/Michael addition.
4. Catalysis and biocatalysis
   Students perform the reduction of ethyl acetoacetate using Baker's yeast.
5. Renewable resources and emerging greener technologies
   Students study renewable feedstock as energy sources to make polymers and other chemicals.
   Students perform an electrochemical synthesis in an ionic liquid.
6. Design of a greener product or process
   Individual students develop and present a proposal for a greener product or process.

lecture, films, and weekly discussions. From a cognitive domain standpoint, lectures targeted GC together with basic chemistry content. More specifically, the course was organized according to five main themes, namely (a) introduction, (b) chemical accounting, (c) chemicals and materials, (d) energy, and (e) applications, which were broken down into chemical concepts and GC concepts. In lecture, one of the key assignments was a collaborative group project on the life cycle assessment of a known green commercial product or process. Each group made a 30 min PowerPoint presentation at the end of the term. In lab, students learned both fundamental skills and science process skills. However, one novel aspect of lab was the "Atom Economy Workshop," which was introduced to study the green synthesis of aspirin where students discussed and compared percent yield to percent atom economy. Student evaluations were reported as very positive.

In 2013, Prescott [54] described an innovative course, one of the first reported in the literature, whose purpose was to teach chemistry fundamentals from a GC perspective to nonmajors. The course, worth three-credit lecture, was offered as part of a novel general education curriculum designated as discovery education. The course, with only nine students, met once a week to meet the needs of the nontraditional student body who were commuters. The course followed a lecture-discussion format and made extensive use of active learning, inquiry pedagogy, and group learning, in which students used metacognition to reflect and take corrective action on their learning deficiencies. GC was integrated with chemistry content. To promote discussion, students were asked to read assigned chapters from Stanley Monahan's online text [55] that was used as a class resource. Class discussions often followed the problem-based learning model where students are posed a question, and are then provided the information and skills required to solve it; students then work toward a collaborative solution through discussion. The problem-based learning model [56, 57] has been shown to increase both the comprehension and retention of new information. Students were assessed on both chemistry and GC content through midterm and final exams, a weekly blog entry, and a semester project that was disseminated by wiki and a public symposium. The blog increased student participation, provided starting points for the next classroom meeting, kept students connected outside the classroom, and helped students discuss and reflect on their knowledge. The final project, either individual or group, integrated general chemistry content knowledge. It was then posted on a class wiki. The course successfully covered most of the important topics studied in a traditional general chemistry course but the author identified three problems. The first problem was that it was not possible to cover all of the text chapters, presumably because of the course discussion/inquiry format. The second problem was that to promote better assessment of fundamental concepts, exams needed to be redesigned with the goal of providing more frequent feedback. The third major problem was that more frequent feedback on student blog entries would help. Students also completed the SENCER-SALG (Student assessment of learning gains) survey that measured gains in student learning. The results were positive, although the student sample was small (n = 6).

In summary, only a few valuable studies have been reported in the literature, demonstrating successfully designed GC courses. One interesting observation is that so many reports have been published demonstrating that GC motivates students to learn chemistry that perhaps a "GC Effect" is at play, combating chemophobia through bias reversal.

6.3.7 Organic GC Pedagogy

In this last section of Part 1 will be reviewed organic pedagogy that could be transferred to other courses. In 2007, Gaddis and Schoffstall [58] published a very well-written paper comparing and contrasting cookbook and inquiry pedagogies in undergraduate organic chemistry laboratories.

According to the authors, cookbook labs suffer from several disadvantages including that they are instructor-centered, emphasize following rote procedures in a step-by-step manner, and verify known results. The authors recommend the format of the guided inquiry laboratory because it allows student freedom without overburdening them in experimental and cognitive design.

In 2014, Graham et al. [59] published how to set up inquiry-based project labs for organic chemistry. Citing that inquiry labs more closely mimic the practice of research, the authors also noted the disadvantages of cookbook labs [60, 61]. Students worked in pairs and after picking an organic synthesis problem, they completed a literature search and designed a greener approach.

Goodwin [62] presented a "green scientific method" as a heuristic to evaluate the pedagogical and green benefits of organic labs through a risk versus reward strategy. This was done by critically examining SDS (safety data sheets) and toxicology data for reagents used and products formed, and by additionally considering the educational benefits of a lab. The evaluation process involved three steps:

1. What was green about the experiment?
2. What was not green?
3. How could the experiment be made greener?

If the educational benefits of an existing lab are low, then the nongreen lab is optimized, replaced, or eliminated. He notes that no organic reaction can ever be 100 % green, and therefore experimental optimizations can only asymptotically approach a green ideal.

The last pedagogy reviewed involves how to set up a green organic chemistry research project [63] in organic chemistry lab. Undergraduate chemistry majors transformed a nongreen organic synthesis to a greener reaction by working through published procedures in the research literature. Students worked in pairs, and they ran both the given nongreen reaction and its optimized counterpart. This guided inquiry project took place over several weeks and was successful at engaging students, allowing independence, creativity, increasing confidence, and appreciating the importance of GC. Furthermore, it helped students understand the link between academics and research.

6.4 Part 2: Sustainable Chemistry Pedagogy: A Historical Approach

Part 2 reviews publications addressing sustainable chemistry, and education for sustainable development (ESD) taking a chronological approach. One reason for this section is to describe the trend over the past 5 years, in which chemistry education journals have shifted emphasis from using the term "green chemistry" to "sustainable chemistry." This section will describe published pedagogy and will summarize current research on ESD.

From a chemistry perspective, sustainable chemistry carries a more global connotation than GC. However, there are several perspectives on the relationship between GC and sustainability. While some take the view that sustainable chemistry is a broader concept that includes GC as a subfield, and also includes energy consumption, waste management, environmental impact, ecology, and human conditions, others regard the two ideas to have merged and imply that they have common goals describing how chemists contribute safer and less hazardous products and processes to sustain society and protect the environment. In other contexts, they are treated as separate fields. Sustainable educators believe that the success of sustainable chemistry depends on exposing science students to GC and sustainability. Moreover, they believe that the pursuit of sustainability is impossible to reach without the chemical industry's participation because chemical products are an integral component of society.

Much of the impetus for sustainable chemistry arises from the UN's declaration of a worldwide Decade of Education for Sustainable Development (DESD) which ran from 2005 through 2014. One of the major goals of DESD was in educating young people in ESD. Hence, a number of research papers were published to address sustainability and DESD.

Another evolving trend regarding ESD was to expand GC and sustainability and integrate them with the social and political sciences to formulate cross-disciplines that address socioscientific issues (SSIs). In order to integrate chemistry education and sustainable development into chemistry education strategies, science teachers must be trained in ESD theory, pedagogy, and practice.

Between 1995 and 2015, 37 papers with sustainability in the title appeared in the JCE, and 10 of these carried both "green chemistry" and "sustainability." The distribution of paper according to topic is given in Table 6.3.

In addition, seven papers appeared in *CERP* from 2012 to 2015, and one in *The Chemical Educator*.

Table 6.3: Sustainability papers published in JCE.

Topic	Number	Years
The general concept of sustainability	5	2008–2013
College courses	7	2013–2014
Organic chemistry experiments	4	2008–2010
Resources for GC and sustainability	1	2013
Book reviews	3	2011–2015
Outreach	3	2014
Conferences	3	2013
Sustainability and education	3	2012–2013
Lab manual for introductory chemistry	1	2013
Life cycle assessment (LCA)	1	2012
Research project	1	2015
Catalytic science	1	1995
Fuel chemistry, biodiesel, biorefinery	3	2013
Lab field trips	1	2015

In *JCE*, the first published paper that carried the concept of sustainability in its title was published by Cusumano [64] in 1995. He wrote about the value of chemical catalysts in spurring economic and environmentally sustainable growth.

Next in 2003, one paper about sustainability was published by *The Chemical Educator*. In this paper, noting the finite abundance of natural resources, Diehlmann et al. [65] urged chemists to contribute to sustainability through ESD and GC education to younger chemists. Moreover, the paper urged the development of “sustainable synthesis optimization rules” to standardize green syntheses techniques.

The next set of papers appeared in 2008 when sustainability was an issue theme in *JCE*. Moore [66] wrote an editorial in which he called for better funding of education and research in order to address worldwide problems such as climate change. In the same issue, Iyere [67] wrote a commentary on the concept of sustainability and explained how chemistry can play an important role in promoting it, in order to solve world hunger and poverty. Pence [68] described sustainability education activities developed by the ACS Committee on Environmental Improvement and commented “that it would be irresponsible for chemistry professors to allow students to graduate without exposing them to issues of sustainability in their classes.”

Also in 2008, Eissen et al. [69] described an open-access online resource that is used to integrate sustainability into organic chemistry lab. In a collaborative project between six German universities, a lab manual was developed with sustainability being the theme of each lab.

Between 2009 and 2011, only two papers, written by Kirchoff [70] and Buntrock [71], addressed sustainability in very general terms. In an editorial, Kirchoff explained how GC can contribute to sustainability and help use the Earth’s resources in a wiser manner.

In 2012, Eiksa and Rauch [72] wrote, in a preface to a series of CERP papers, that sustainable development is a political aim. Their perspective emphasized producing sustainably literate citizens by making sustainability in research and industry a top goal and seemed to shift the focus of GC education to an ESD viewpoint. Moreover, the authors indicated that chemical education should produce students at the secondary and college levels who both understand chemical technology and participate actively in society through debate. To implement this plan, the authors indicated that students should achieve chemistry content knowledge but equally study sustainability issues and understand important issues pertaining to industry and the environment. They presented several examples of GC SSIs such as plastics or organic syntheses.

In 2012, five papers were published, two of which were general narratives on sustainability; each of these papers will be described in turn. Noting that sustainability is a multidisciplinary big idea and that ACS guidelines promote preparing students who are aware of global and environmental issues, Fisher [73, 74] proposed increased activity on the part of chemical educators to incorporate sustainability into their programs citing the availability of many good resources. Zoller [75] recommended that chemistry education undergo a paradigm shift to a scientific, environmental, and technological education that incorporates science literacy for sustainability. Another paper [76] addressed sustainable organic chemistry by providing experimental and mechanistic details for an aqueous multicomponent Petasis Borono–Mannich reaction. Pohl et al. [77] described several sustainable chemistry experiments for the undergraduate organic laboratory curriculum, where various plant oils underwent saponification and transesterification reactions to create soaps and fuels. Vervaeke [78] described life cycle assessments for evaluating product impact in the environment.

In 2013, much of a *JCE* issue was devoted to ConfChem sustainability papers, noting that “Education is the key to a sustainable future.” Several papers [79, 80] approached sustainability and ESD from the United Nations (UN) and general information viewpoints, while one paper [81] addressed sustainable energy. Young and Peoples [82] described educational materials for both educators and students developed for ESD by the ACS and the Green Chemistry Institute. Chiu [83] wrote about globalization and international standards for chemistry education. These standards are designed to cultivate student skills in seven perspectives across the globe: content knowledge, contextual learning, communication, critical thinking and inquiry, interdisciplinary connections, creativity, and commitment to chemistry.

In order to introduce the sustainability of fuels to science college students, Wallington [84] provided an inquiry approach that can be used in science education lectures at the high school or undergraduate level. First, students were provided with an understanding of the elemental composition of future fuels. They then conducted a systematic search through the periodic table and discovered that viable fuels contain only three elements: C, H, and O. Then using knowledge of the properties that make fuels mobile and inexpensive, students inferred what types of fuels would be useful for future transportation.

6.4.1 Courses

Published papers about courses on chemistry sustainability are reviewed in this section.

Burmeister and Eilks [85] described how a novel lesson plan on plastics can be implemented and assessed in secondary school chemistry teaching as ESD, by using a sociocritical and problem-oriented approach to chemistry teaching. The authors used a Chem. Ed.-ESD model as recommended by UNESCO (2006) that

1. is embedded throughout entire chemistry curriculum and is interdisciplinary;
2. imbues the chemistry curriculum with values reflecting the goals of ESD;
3. advances and develops critical thinking and problem-solving skills required to meet the challenges of sustainable development;
4. promotes ESD utilizing educational strategic methods in debate, art, and the written word;
5. invites learners to participate in decision-making;
6. integrates ESD learning into daily and personal and professional life.

In another paper, the lesson plan employed by Eilks and Ralle [86] involved a Participatory Action Research (PAR) model in science education. PAR involves a collaboration between teachers (practice) and researchers (theory) who study science pedagogy through a cyclic process that involves designing, testing, modifying, and reflecting on research plans, resulting in novel teaching strategies and materials. The starting point involves an authentic current events' socioscientific controversy designed to engage and motivate students to learn science content. Students then debate the issue and reflect on their learning. The five steps in PAR are summarized below. Students

1. analyze the problem;
2. perform chemical lab work to elaborate the problem;
3. apply the socioscientific dimension;
4. debate, discuss, and evaluate different viewpoints; and
5. engage in metacognitive reflection.

Experiences and feedback from teachers and students based on the cyclical development by PAR are discussed.

Karpudewan et al. [87] described a novel ESD-GC pedagogy successfully used in both high school science course, and in a methods course for pre-service teachers in Malaysia. The authors stated that their work helped integrate GC into chemistry and chemistry education curricula, and may serve as an instructional model for GC education in developing countries. However, in a broader sense it is clear that their work could be extended to any similar methods course. The authors primarily used a laboratory approach, by introducing a GC lab manual containing 27 experiments devoted to the pre-service teacher and secondary school chemistry curricula. Training and pretesting experiments were provided through workshops, in order to develop an effective and student-centered approach. Prelab questions and discussions preceded each experiment. In addition, preservice teachers learned how to convert a traditional experiment into a GC experiment. Post-lab discussions of chemistry content, GC implications, and societal/economic impacts followed each experiment. Two important sustainable concepts, natural resource accounting and life cycle analysis, were addressed, within the context of the environmental problems facing Malaysia, namely logging and pollution. In addition, students performed an experiment to produce biodiesel from palm oil, used as cooking oil, and also determined the amount of soot generated during burning and measured its heat of combustion as a lab exercise. Feedback received through surveys and assessments indicated that preservice teachers responded in a very positive way to GC principles, but found that some of the experiments could be more clearly written. For the actual secondary students involved in the study, results showed that both student motivation and content achievement significantly increased after using the GC lab curriculum. It is worthy to note that although Malaysia's mean PISA scores (reading, mathematics, science) are in the lower third, they are introducing GC into their chemistry curricula rather than solely concentrating on chemistry content.

In another paper, Marteel-Parrish [88] wrote about the evolution of an elective GC course titled "Toward the Greening of Our Minds: Green and Sustainable Chemistry." The course was offered to both chemistry and biology majors and minors between 2005 and 2014 at Washington College. In the course, GC instruction evolved through an iterative process. Moreover, students elaborated on their understanding of GC content by applying its tenets to real-world problems such as evaluating reaction greenness, chemical safety and sustainability, using a green decision tree, and promoting sustainability in underdeveloped countries. Students participated through daily critical discussion and writing. Student exercises involved, for example, comparing and contrasting the relative greenness of traditional commercial products and processes versus green ones. Course iterations over the years addressed these modifications and improvements:

1. GC metrics introduced included LCA, percent atom economy, and E-factor. They were applied to organic reactions such as cycloadditions, electrophilic aromatic substitution, and Suzuki coupling.
2. To promote understanding of business and green decision making, students evaluated political, societal, and business drivers in the success of GC products and processes. Students studied examples and created case studies of the successful development and implementation of green commercial processes leading to viable products. Students took into account these factors: cost to industry, the environment, health/safety, and the strengths and weaknesses of the greener product or process. Finally, students critically evaluated the outcomes of their case study using criteria in economics, production, and public image.
3. Other course modifications improved student understanding of GC in developing countries, and health and safety. Student feedback was very positive.

A novel course [89] "Green Chemistry and Sustainability" was offered for undergraduate Honors College multidisciplinary science majors and nonmajors at Creighton University. Students studied GC, surveyed the scientific literature, developed technical writing skills, and presented oral group research proposals. The novel course design was based on both environmental chemistry and the 12 principles of GC. More specifically, the three environmental themes and their GC components were as follows: (a) energy: catalysts, energy efficiency, and renewable feedstocks; (b) pollution and waste prevention: pollution, waste, atom economy, avoiding chemical derivatives, and designing for degradation and less hazardous syntheses; and (c) safety: for better products and solvents, and increased accident prevention. Students went beyond the cognitive domain and were exposed to the affective domain in which they thought critically about how chemistry can affect the Earth and its inhabitants and is linked to the sustainable earth.

Although Kovacs [90] published the next article in 2013, it addressed a novel GC course titled "Green Chemistry and Industrial Processes" that was first offered at Grand Valley State University in 2006. It was designed to bridge the gap between academia and industry by applying the principles of GC to real life through a partnership with several local businesses. The course was designed based on the results of a student designed survey of about 1,600 students of which about 70 % responded positively to wanting to learn more about GC.

Haack et al. [91] published a 2013 description of a general education science course called "Chemistry of Sustainability" that was designed and taught through faculty collaboration and team teaching. Chemistry content material supported teaching GC and sustainable issues using both a case study and problem-solving approach that utilized faculty research expertise in specific areas such as renewable energy, sustainable consumer products, bioplastics, clean water, and nanoscience. The project positively impacted everyone involved, including students and faculty team members, who developed new education materials and new perspectives. One benefit to the chemistry curriculum was that GC and sustainability were infused into more advanced undergraduate and graduate courses.

Cummings [92] provided information about "Solar Energy: A Chemistry Course on Sustainability for General Science Education and Quantitative Reasoning," which was offered as part of the liberal arts curriculum and environmental studies minor. The course emphasized the crucial role that chemistry plays in solving energy problems by finding new and clean energy sources thereby preventing pollution. The course addressed the problems caused by fossil fuel combustion and then explored clean alternatives such as solar energy, including solar radiation, combustion, greenhouse gases, ethanol, photovoltaics, water electrolysis, fuel cells, hydrogen storage, and batteries. In order to understand the chemistry associated with fossil fuels and solar energy, content chemistry in the following areas was taught: reaction stoichiometry, molecular structure, thermochemistry, catalysis, energy quantization, and electrochemistry. Students utilized critical thinking skills and quantitative reasoning. Moreover, the liberal arts nature of the course fostered interdisciplinary connections with economics, politics, ecology, and human health. Students were assessed on their participation in discussions, lab assignments, and performance on pre- and posttests on these areas: solar energy policy, applied solar energy, and solar energy. Students showed significant gains. Moreover, students rated the course with good reviews.

Brinkert et al. [93] cited the United Nations Rio Declaration on Environment and Development to justify a graduate-level course on solar energy and sustainable development open to students from both nonscience and science backgrounds. The course may serve as a model utilizing novel content and pedagogy for other academics to introduce a similar course on sustainability into other chemistry programs. The title of the course "Perspectives on Solar Energy from Science, Industry and Policy" was a collaborative effort combined with the Center for Sustainable Development and the Department of Chemistry, Ångstrom Laboratories, at the Uppsala University, Sweden, in spring 2012. One goal was to design and implement a cross-disciplinary university course to train both academic and nonacademics from a variety of backgrounds, including nonscientists, to meet the growing demands of ESD across the university curriculum. Everyone, including nonacademics, could apply to participate in the course even though they lacked background knowledge on science and solar energy. The selected student cohort represented three distinct groups: social science, natural science, and solar energy professionals who were trained in natural science.

The course provided a general review of solar energy through the balanced views of science, policy making, and industry. It was further organized around these components delivered as modules: politics, humanitarian, industrial, and solar energy. The 3-week course was limited to 16 participants who were selected based on their different educational backgrounds, cultural diversity, and motivation. The course was deployed in a format involving lecture followed by in-depth discussion; it also utilized guest lecturers. Students then completed additional assignments such as labs. Before each assignment, students prepared short assignments that complemented their research background.

A course named "Organic Synthesis Techniques" was described by Dicks and Batey [94], which was an advanced organic lecture-lab course whose main goal was to introduce green and sustainable chemistry principles to undergraduates through the study of catalytic methods. New experiments were designed to investigate these catalytic methods: phase-transfer catalysis, organocatalysis, Lewis and Brønsted acid catalysis, and transition metal catalysis. In lectures, students studied the principles of GC, and compared and contrasted catalyzed syntheses to traditional ones through industrial case studies. In addition to catalysts, in lab, students also investigated these types of reaction solvents: solvent free, water, greener organic solvents, ionic liquids, and supercritical CO_2.

6.4.2 Other Papers

Fisher's paper [95] published in 2012 supports the idea of including the global issue of sustainability as part of a professional curriculum reform in chemistry because it provides a richer context and more well-rounded liberal arts educational experience, meets the United Nations declaration calling for a decade of sustainable development, and last of all, meets an ACS guideline that chemistry majors should be aware of the role of chemistry in contemporary societal and global issues. Moreover, according to Sherman [96], the focus of ESD in higher education should address these areas of (a) determining prescribed practices; (b) campus operations; and (c) instituting new academic programs.

Through a "perspective," Burmeister et al. [97] described how chemistry education can incorporate the UN's DESD plans into their educational programs. The article began by justifying ESD across the three domains ecology, economics and society. The authors recognized the important role that GCE played in sustainability and cited many examples, such as the ozone hole problem, that can be used as pedagogical case studies that lead to positive changes in student knowledge and understanding. They suggested four basic pedagogical models ranging from narrow to wide involvement with regard to their respective academic communities. These models that can be combined and have the capacity to infuse sustainable development into formalized chemistry education curricula are given in Table 6.4. The authors believed that models 3 and 4 have the most potential for successfully integrating ESD with chemical education because students will learn about and learn to contribute to sustainable development. However, all four ESD proposed models require much more than modifications to curricula, but instead require redirection and redesigning curricula across interdisciplinary science fields, and extending them to include society, economics, and ecosystems.

A Finnish article [101] offers an in-depth review of the status of ESD in chemistry (ESD-Chem) by analyzing current models for their relative strengths and weaknesses, and offering three distinct pedagogical models for future practice in Finland. Although the goals of ESD-Chem are straightforward, namely empowering citizens, consumers, and educators to act on the levels of the individual, the community, and the ecosystem for a sustainable world, implementing it in the classroom is a very complicated enterprise. The complications relate to more than integrating the concepts of sustainability and GC and arise when incorporating SSI into the classroom. Therefore, this Finnish model may not be a viable pedagogy for US secondary schools that are focusing on attempting to overcome chemistry content deficiencies, whereas Finnish PISA scores are in the top tier. After analyzing current models of ESD-Chem pedagogy, they propose these new models:

1. A holistic ESD-basic chemistry pedagogy offers these attributes: interdisciplinary, student centered, inquiry based, with social interaction to promote socioscientific argumentation practices by examining SSIs and societal cooperation with stakeholders.

Table 6.4: Pedagogical models for GCE and ESD.

Model 1: Apply GC principles to science education lab work
This model applies Anastas and Warner's [98] GC philosophy to: 1. Reduce quantities of reagents by shifting to the micro-scale 2. Substitute less dangerous substances 3. Employ catalysts to lower energy requirements 4. Produce less waste 5. Use continuous reflection to cement GC student learning
Model 2: Integrate sustainability with chemistry content chemistry education
This approach introduces GC and sustainability together with chemistry content making the instruction relevant through using successful examples involving energy, renewable fuels, industry, and pharmaceuticals. By introducing GC early on, students better assimilate the subject and become more science literate.
Model 3: Using controversial current event issues regarding sustainability
By using controversial issues, teachers can stimulate student interest in GC and sustainability. However, this model utilizes explicit chemistry content knowledge, as well as focusing on how chemistry impacts society. For example, controversies related to biofuel use [99, 100] could be used as main topic.
Model 4: Integrating chemical content knowledge with ESD across an institution
Model 4 was developed with elementary and secondary schools in mind. In it, chemistry instruction together with ESD becomes the institutional goal and the common thread of all school curricula and activities. Students become active citizens who contribute to the sustainability of their educational community.

2. The second model's dominant characteristic shows the teacher how to implement a student-centered strategy.
3. In the third model, a three-phase strategy implements SSI instruction through empowering students.

Overton and Randle [102] published a 2015 paper related to teaching sustainable chemistry to first year chemistry undergraduates. One purpose of the course was to train chemistry students in sustainable development before they enter industrial careers. In particular, students used dynamic problem-based learning, a constructivist approach, to solve real-world problems such as designing sustainable residential villages subject to a variety of constraints such as cost, environmental impact, and trading partly limited to the local economy. Another interesting problem was deciding whether biodiesel or bioethanol was more cost effective as a fuel for a fleet of 42 buses. Students presented written and oral group reports.

Acknowledgements

I thank Jessica Crowley for performing part of the literature search used in this paper.

References

[1] National Science Board. National Action Plan for Addressing the Critical Needs of the US Science, Technology, Engineering, and Mathematics Education System, 2007, http://www.nsf.gov/nsb/documents/2007/stem_action.pdf (accessed September 2015).
[2] National Science Board. Revisiting the STEM Workforce: A Companion to Science and Engineering Indicators 2014, NSB-2015-10, February 4, 2015, http://www.nsf.gov/pubs/2015/nsb201510/nsb201510.pdf (accessed September 2015).
[3] Rothwell, J. The Hidden STEM Economy. (Washington, DC: Brookings Institution, 2013). http://www.brookings.edu/research/reports/2013/06/10-stem-economy-rothwell; Executive Office of the President. President's Council of Advisors on Science and Technology (PCAST) Memo of September 2014. (Washington, DC: Executive Office of the President, 2014) (accessed July 2015).
[4] Special Report: Committee on Professional Training Spring 2008 American Chemical Society. http://www.acs.org/content/dam/acsorg/about/governance/committees/training/reports/cptreports/phd-programs-in-chemistry-survey-report-2008.pdf (accessed September 2015).
[5] 2014 Salaries and Employment. American Chemical Society. CEN ACS.org http://www.acs.org/content/dam/acsorg/careers/salaries/cen-salary-article.pdf (accessed September 2015).
[6] US Department of Education Report titled Education for Global Leadership, STEM http://www2.ed.gov/about/overview/budget/budget14/crosscuttingissues/stem.pdf (accessed September 2015).
[7] OECD. http://www.oecd.org/pisa/ (accessed September 2015).
[8] OECD/Pisa website. http://www.oecd.org/pisa/keyfindings/pisa-2012-results.htm (accessed September 2015).
[9] Love Canal American Chemical Society History of Green Chemistry. http://www.acs.org/content/acs/en/greenchemistry/what-is-green-chemistry/history-of-green-chemistry.html (accessed September 2015).
[10] Kaiser, D. *Pedagogy and the Practice of Science: Historical and Contemporary Perspectives*. The MIT Press: Cambridge and London, 2005.
[11] Donovan, M.; Suzanne, B.; John D. (eds.). *How Students Learn Science in the Classroom, National Research Council Committee on How People Learn.* National Academies Press: Washington, DC, 2005.
[12] Bodner, G. Constructivism, a theory of knowledge. *J Chem Educ* **1986**, 63 (10), 873–878.
[13] Bodner, G.; Klobuchar, M.; Geelan, D. The many forms of constructivism. *J Chem Educ* **2001**, *78* (8), 1107–1134.
[14] Eick, C.; Meadows, L.; Balkcom, R. Breaking into Inquiry. *Sci Teach* **2005**, *72* (7), 49–53.
[15] Wilcox, J.; Kruse, J. W.; Clough, M. P. Teaching science through inquiry: Seven common myths about this time honored approach. *Sci Teach* **2015**, *82* (6), 62–67.
[16] Shiland, T. W. Constructivism, the implications for laboratory work. *J Chem Educ* **1999**, *76* (1), 107.
[17] Fink, L. D. *Creating Significant Learning Experiences*. Jossey-Bass: San Francisco, CA, 2003.
[18] Levy, I. J.; Haack, J. A.; Hutchison, J. E.; Kirchhoff, M. Going Green. Lecture assignments and lab experiences for the college curriculum. *J Chem Educ* **2005**, *82* (7), 974.
[19] Andraos, J.; Dicks, A. P. Green chemistry teaching in higher education; a review of effective practices. *Chem Educ Res Pract* **2012**, *13*, 69–70.

[20] Journal of Chemical Education. http://pubs.acs.org/journal/jceda8 (accessed September 2015).
[21] Chemical Education Research and Practice. http://pubs.rsc.org/en/journals/journalissues/rp#!recentarticles&adv (accessed September 2015).
[22] National Science Teachers Association. http://www.nsta.org/publications/digitaljournals.aspx (accessed September 2015).
[23] Marasco, C. A. The ivory tower goes green. *Chem Eng News* **2008**, *86* (36), 64–66.
[24] American Chemical Society. http://www.acs.org/content/acs/en/greenchemistry/students-educators/academicprograms.html (accessed September 2015).
[25] Search for ACS Approved Programs. https://webapplications.acs.org/Applications/CPTASL/app_list_search.cfm?CFID=27536&CFTOKEN=640a62017b2d1051-FCC5C897-DE32-D25F-9DCB-CA7C2081835C&jsessionid=5168894E856D01645733472ABB05A66A.cfusion (accessed September 2015).
[26] Beyond Benign: Green Chemistry Education. http://www.beyondbenign.org/ (accessed September 2015).
[27] Celebrating Chemistry/American Chemical Society. http://www.acs.org/content/acs/en/education/outreach/celebrating-chemistry-editions.html (accessed September 2015).
[28] Green Chemistry/American Chemical Society. http://www.acs.org/content/acs/en/about/makedonation/greenchemistry.html (accessed September 2015).
[29] NGSS. http://www.nextgenscience.org/search/node/sustainability (accessed September, 2015).
[30] Sullivan, M. Career of the month: an interview with green product chemist Andy Chen. *Sci Teach* **2007**, *74* (8), 87.
[31] Roy, K. Greener is cleaner and safer. *Science Scope* **2005**, *28* (6), 50–52.
[32] Roy, K. Clean and Green. *Science Teach* **2012**, *79* (5), 68–69.
[33] NSTA Web News Digest. Scope on safety. Greener is cleaner and safer. http://www.nsta.org/publications/news/story.aspx?id=50316. (accessed Sept. 9, 2016).
[34] Clary, R.; Wandersee, The green root beer laboratory. J. *Sci Teach* **2010**, *77* (2), 24–28.
[35] Palliser, J. Green science. Green beauty. *Sci Scope* **2010**, *34* (4), 8–11.
[36] Mandler, D.; Mamlok-Naaman, R.; Blonder, R.; Yayon, M.; Hofstein, A. High school chemistry teaching through environmentally oriented curricula. *Chem Educ Res Pract* **2012**, *13*, 80–92.
[37] Cacciatore, K. L.; Hanna, S. Teaching lab report writing through inquiry. A green chemistry stoichiometry experiment for general chemistry. *J Chem Educ* **2006**, *83* (7), 109–1041.
[38] Cacciatore, K. L.; Amado, J.; Evans, J. J. Connecting solubility, equilibrium, and periodicity in a green inquiry experiment for the general chemistry lab. *J Chem Educ* 2008, *85* (2), 251–253.
[39] Donovan, S.; Bransford, J. (eds.). *How Students Learn: History, Mathematics and Science in the Classroom*. National Academies Press: Washington, DC, 2005.
[40] Singer, S. R.; Hilton, M. L.; Schweingruber, H. A. (eds.). *America's Lab Report: Investigations in High School Science*. National Academies Press: Washington, DC, 2005.
[41] Bransford, J.; Brown, A. L.; Cocking, L. L. (eds.) *How People Learn: Brain, Mind, Experience and School*. National Academies Press: Washington, DC, 2000.
[42] Song, Y. M.; Wang, Y. C.; Geng, Z. Y. Some exercises reflecting green chemistry concepts. *J Chem Educ* **2004**, *81* (5), 691–692.
[43] Klingshirn, M. A.; Wyatt A. F.; Hanson R. M.; Spessard, G. O. Determination of the formula of a hydrate: A greener alternative. *J Chem Educ* **2008**, *85* (6), 819–821.
[44] Wellman, W. E.; Noble, M. E. Out of the blue. *J Chem Educ* **2003**, *80* (5), 537–540.
[45] McCarthy, S. M.; Gordon-Wylie, S. W. A greener approach to measuring colligative properties. *J Chem Educ* **2005**, *82* (1), 116–119.
[46] Jansen, M. P. The cost of converting a gasoline powered vehicle to propane. A practical review problem for senior high school or introductory chemistry. *J Chem Educ* **2000**, *77* (12), 1578.
[47] Greener education for chemists. http://greenchem.uoregon.edu/gems.html. (accessed Sept. 8, 2016).

[48] Hjeresen, D. L.; Schutt, D. L.; Boese, J. M. Green chemistry and education. *J Chem Educ* **2000,** *77* (12), 1543–1547.
[49] Cann, M. C.; Dickneider, T. A. Infusing the chemistry curriculum with green chemistry using real world examples, web modules, and atom economy in organic chemistry courses. *J Chem Educ* **2004,** *81* (7), 977–980.
[50] Braun, B.; Charney, R.; Clarens, A.; Farrugia, J.; Kitchens, C.; Lisowski, C.; Naistat, D.; O'Neil, A. Completing our education: Green chemistry in the curriculum. *J Chem Educ* **2006,** *83* (8), 1126–1129.
[51] Collins, T. J. Introducing green chemistry in teaching and research. *J Chem Educ* ***1995,*** *72* (11), 965–966.
[52] Cann, M. C. Bringing state of the art, applied, novel, green green chemistry to the classroom by employing the Presidential Green Chemistry Challenge Awards. *J Chem Educ* **1999,** *76* (12), 1639–1641.
[53] Manchanayakage, R. Designing and incorporating green chemistry courses at a liberal arts college to increase students' awareness and interdisciplinary collaborative work. *J Chem Educ* **2013,** *90* (9), 1167–1171.
[54] Prescott, S. Designing and teaching a general chemistry laboratory course to nonmajors using a green chemistry approach. *J Chem Educ* **2013,** *90* (9), 423–428.
[55] Manahan, S. E. *Green Chemistry and the Ten Commandments of Sustainability*; 2nd ed., ChemChar Research, Inc.: Columbia, MO, 2006.
[56] Schmidt, H. G.; Rotgans, J. I.; Yew, E. H. The process of problem based-learning: what works and why. *J Med Educ* **2011,** *45,* 792–806.
[57] Dolmans, D.; Schmidt, H. The advantages of problem-based curricula. *Postgrad Med J* ***1996,*** *72,* 535–538.
[58] Gaddis, B.; Schoffstall, A. M. Incorporating Guided-Inquiry Learning into the Organic Chemistry Laboratory. *J Chem Educ* **2007**, *84* (5), 848–851.
[59] Graham, K. J.; Jones, T. N.; Schaller, C. P.; McIntee, E. J. Implementing a student-designed green chemistry laboratory project in organic chemistry. *J Chem Educ.* **2014,** *91* (11), 1895–1900.
[60] Monteyne, K.; Cracolice, M. S. What's wrong with cookbooks? A reply to Ault, Mark S. *J Chem Educ* **2004,** *81* (11), 1559–1560.
[61] Horowitz, G. What's wrong with cookbooks? *J Chem Educ* **2008,** *85* (1), 346–353.
[62] Goodwin, T. E. An asymptotic approach to the development of a green organic chemistry laboratory. *J Chem Educ* **2004,** *81* (8), 1187–1190.
[63] McKenzie, L. C.; Huffman, L. M.; Hutchison, J. E.; Rogers, C. E.; Goodwin, T. E.; Spessard, G. O. Greener solutions for the organic chemistry teaching lab: Exploring the advantages of alternative reaction media. *J Chem Educ* **2009,** *86* (4), 488–489.
[64] Cusumano, J. A. Environmentally sustainable growth in the 21st century. *J Chem Educ* **1995,** *72* (11), 959.
[65] Diehlmann, A.; Kreisel, G.; Gorges, R. Contribution to "Developing Sustainability" in Chemical Education. *J Chem Educ* **2003,** *8* (2), 102–106.
[66] Moore, J. W. Sustainability. *J Chem Educ* **2008,** *85* (1), 1595.
[67] Iyere, P. A. Chemistry in Sustainable Development and global environment. *J Chem Educ* **2008** *85* (12), 1604–1606.
[68] Pence, L. E. Sustainability education and the ACS committee on environmental improvement. *J Chem Educ* **2008,** *85* (12), 1695.
[69] Eissen, M.; Bahadir, M.; König, B.; Ranke, J. Developing and disseminating NOP: an online, open-access, organic teaching resource to integrate sustainability concepts in the laboratory. *J Chem Educ* **2008,** *85* (7), 1000.
[70] Kirchoff, M. M. Education for a Sustainable Future. J Chem Educ **2010,** 87 (2), 121.

[71] Buntrock, R. E. Book review of image and reality, Kekule, Kopp and the scientific imagination, *J. Chem. Educ* **2011,** *88* (10), 1358.
[72] Eilks, I.; Rauch, F. Sustainable development and green chemistry in chemistry education. *Chem Educ Res Pract* **2012,** *13*, 57–58.
[73] Fisher, M. A. Chemistry and the challenge of sustainability. J Chem Educ **2012,** 89 (*2*), 179–180.
[74] American Chemical Society Web Site for Standards, Guidelines, and ACS Approval Program. http://portal.acs.org/portal/PublicWebSite/education/policies/index.htm (accessed July, 2015).
[75] Zoller, U. Science education for global sustainability. What is necessary for teaching, learning, and assessment strategies. *J Chem Educ* **2012,** *89* (3), 297–300.
[76] Candeias, N. R.; Paterna, R.; Cal, P. M. S. D.; Góis, P. M. P. A sustainable protocol for the aqueous multicomponent Petasis Borono-Mannich reaction. *J Chem Educ* **2012,** *89* (6), 799–802.
[77] Pohl, N. L. B.; Streff, J. M.; Brokman, S. Evaluating sustainability: Soap versus biodiesel production from plant oils. *J Chem Educ* **2012,** *89* (8), 1053–1056.
[78] Vervaeke, M. Life cycle assessment software for product and process sustainability analysis. *J Chem Educ* **2012,** *89 (7),* 884–890.
[79] Belford, R. E.; Bastin, L. D. Using a progressive paper to develop students' writing skills. *J Chem Educ* **2013,** *90* (4), 508–509.
[80] Pence, L. E.; Kirchoff, M. M. ConfChem Conference on educating the next generation: Green and sustainable chemistry - Green chemistry and sustainability through the American Chemical Society Education Division and Committee on Environmental Improvement. *J Chem Educ* **2013,** *90* (4), 510–512.
[81] Abruna, H. D. Energy in the age of sustainability. *J Chem Educ* **2013,** *90* (11), 1411–1413.
[82] Young, J. L; Peoples, R. ConfChem Conference on educating the next generation: Green and sustainable chemistry - Education resources from the ACS Green Chemistry Institute. *J Chem Educ* **2013,** *90* (4), 513–514.
[83] Chiu, M. H. ConfChem on a Virtual Colloquium to sustain and celebrate IYC initiatives in global chemical education. Sustainability and globalization of chemical education. *J Chem Educ* **2013,** *90* (11), 1564–1566.
[84] Wallington, Timothy, J.; Andersen, James E,; Siegel, Donald J.;Tamor, Michael, A.; Mueller, Sherry A.; Winkler, Sandra, L.; Nielsen, Ole J. Sustainable mobilty, future fuels, and the periodic table. *J Chem Educ* **2013,** 90 (11), 440–445.
[85] Burmeister, M.; Eilks, I. An example of learning about plastics and their evaluation as a contribution to Education for Sustainable Development in secondary school chemistry teaching. *Chem Educ Res Pract* **2012,** 13, 93–102.
[86] Eilks, I.; Ralle, B. Participatory Action Research in Chemical Education. http://www.idn.uni-bremen.de/chemiedidaktik/material/Symp%202002%20Eilks%20 Ralle %20PAR.PDF (accessed Dec. 19, 2016).
[87] Karpudewan, M.; Ismail, Z.; Roth, W. Ensuring sustainability of tomorrow through green chemistry integrated with sustainable development concepts. Chem Educ Res Pract **2012,** 13, 120–127.
[88] Marteel-Parrish, A. E. Teaching green and sustainable chemistry: A revised one-semester course based on inspirations and challenges. *J Chem Educ* **2014,** 91 (7), 1084–1086.
[89] Gross, E. M. Green chemistry and sustainability: An undergraduate course for science and nonscience majors. *J Chem Educ* **2013,** 90 (4), 429–431.
[90] Kovacs, D. G. ConfChem Conference on educating the next generation: Green and sustainable chemistry - Teaching green chemistry. *J Chem Educ* **2013,** 90 (4), 517–518.
[91] Haack, J. A.; Berglund, J. A.; Hutchison, J. E.; Johnson, D. W.; Lonergan, M. C.; Tyler, D. R. ConfChem on educating the Next Generation: Green and sustainable chemistry - Chemistry of sustainability: A general education science course enhancing students, facultyand institutional programming. *J Chem Educ* **2013,** *90* (4), 515–516.

[92] Cummings, S. D. ConfChem on educating the Next Generation: Green and sustainable chemistry - Solar energy: A chemistry course on sustainability for general science education and quantitative reasoning. *J Chem Educ* **2013,** *90* (4), 523–524.
[93] Brinkert, K.; Pullen, S. SolEn for a sustainable future: Developing and teaching a multidisciplinary course on solar energy to further sustainable education in chemistry. *J Chem Educ* **2014,** *91 (10)*, 1569–1573.
[94] Dicks, A. P.; Batey, R. A. ConfChem on educating the next generation: Green and sustainable chemistry - Greening the organic curriculum: Development of an undergraduate catalytic chemistry course. *J Chem Educ* **2013,** *90* (4), 519–520.
[95] Fisher, M. A. Chemistry and the challenge of sustainability. *J Chem Educ* **2012,** *89* (2), 179–180.
[96] Sherman, D. J. Sustainability. What's the big idea? *Sustainability* **2008,** *1* (3), 188–195.
[97] Burmeister, M.; Rauch, F.; Eilks, I. Sustainable development and green chemistry in chemistry education perspectives: Education for sustainable development (ESD) and chemistry education. *Chem Educ Res Pract* **2012,** *13,* 59–68.
[98] Anastas P. T.; Warner J. C. *Green Chemistry: Theory and Practice.* Oxford University Press: New York, 1998.
[99] Eilks, I. A social critical and problem oriented approach to chemistry teaching and students' first views on it. *Chem Educ Res Pract* **2002,** *3,* 67–75.
[100] Feierabend, T.; Eilks, I. Teaching the societal dimension of chemistry using a social-critical and problem-oriented lesson plan based on bioethanol usage. *J Chem Educ* **2011,** *88* (9), 1250–1256.
[101] Juntunen, M. K.; Aksela, M. K. Education for sustainable development in chemistry - challenges, possibilities, and pedagogical models in Finland and elsewhere. *Chem Educ Res Pract* **2014,** *15,* 488–500.
[102] Overton, T. L.; Randles, C. A. Beyond problem-based learning: Using dynamic PBL in chemistry. *Chem Educ Res Pract* **2015,** *16,* 251–259.

Matthew J. Mio

7 How the Principles of Green Chemistry Changed the Way Organic Chemistry Labs Are Taught at the University of Detroit Mercy

Abstract: Many logistic and instructional changes followed the incorporation of the 12 principles of green chemistry into organic chemistry laboratory courses at the University of Detroit Mercy. Over the last decade, institutional limitations have been turned into green chemical strengths in many areas, including integration of atom economy metrics into learning outcomes, replacing overly toxic equipment and reagents, and modifying matters of reaction scale and type.

Keywords: 12 principles of green chemistry, organic chemistry, atom economy metrics, learning outcomes

7.1 Introduction

Over 10 years ago, a reexamination of the organic chemistry curriculum at the University of Detroit Mercy was embarked upon. The need for such a review was multifaceted. First, rising enrollments encouraged instructors to scrutinize material and equipment usage over time from the perspective of both student safety and course objectives. Increasing ranks of students prepared to study organic chemistry for their preprofessional or scientific training raises questions about the proper number of students to enroll in a lab. This judgment, in turn, engenders discussion about student safety with regard to volatile reagents/solvents, as well as appropriate course learning objectives. The University of Detroit Mercy has utilized the ideologies espoused by the Lab Safety Institute [1], whose work and recommendations are tied to current research in the field of safe lab instruction. In addition, a move toward institutional program review by way of internal self-study forced many labs at the University of Detroit Mercy to reevaluate the goals each course purported to help students achieve.

Second, financial resources at the university were limited, and therefore, economical solutions would be required. While this is no doubt a planning aspect shared by many institutions of higher learning, the University of Detroit Mercy organic chemical lab courses would be bound by the teaching space assigned to them. The space in question, a garden-level laboratory of approximately 2,000 square feet, had several unique characteristics that would need to be kept in mind throughout the course reform. For example, there were only seven fume hoods in the lab, all abutting the outer wall. None of the four long workbenches would have dedicated hood space. Each workstation could accommodate two students working as a pair, but no more than two pairs per bench side would be possible without severely impeding movement in the lab, and therefore, affecting safety. Incidentally, this would set the upper limit of student enrollment for any 3 h lab period to 32 total, even though the recommended number of enrollees would be set lower by square footage standards. There was no house vacuum, house air, or house gases delivered to the benchtop in this lab, and one of the hoods would be used for satellite storage of liquid organic waste.

Third, and most importantly, a call to incorporate the 12 principles of green chemistry into the learning outcomes of requisite courses was set to be heeded. Indeed, the primary impetus for curricular change in the University of Detroit Mercy's organic chemistry lab courses was a need to integrate value into students' instruction. After much dialogue on how to cast the degree outcomes for the Department of Chemistry and Biochemistry, the following four "pillars" were set: scientific method, technical presentation, instrumentation, and social and professional responsibility. The last of these learning outcome groups was to encompass awareness of ethical issues in the chemical sciences, critical reading of the chemical literature, and green chemistry concepts and methods. These objectives align well with both the educational purpose of the American Chemical Society's (ACS) Green Chemistry Institute [2] and the current ACS strategic plan [3]. In fact, it is a main goal of the ACS to improve education by fostering "the development of the most innovative, relevant, and effective chemistry education in the world" [3]. Key to achieving this goal is the incorporation of the theory and application of the 12 principles of green chemistry [4] into higher education instruction. This directive feeds into another of the main goals of the ACS strategic plan, and that of communication of chemistry's value. This goal is summarized by helping chemists communicating chemistry's vital role in addressing the world's challenges to the public and policymakers.

All in all, these chief factors converged with a complete reimagining of three organic chemistry laboratory courses at the University of Detroit Mercy: CHM 2250, a one-credit introductory lab for all science majors; CHM 2260, a one-credit second-semester lab for biology majors; and CHM 2300, a two-credit advanced synthesis lab for chemistry and biochemistry majors.

7.2 Green Chemistry Principles Affect Course Learning Outcomes

Looking back, the most profound change that resulted from the incorporation of the theory and application of green chemistry principles into the University of Detroit Mercy's organic chemistry lab courses was the inclusion of green chemistry metrics. Popular generalizations of the 12 principles divide the spirit of their responsibility into four major groups: energy usage, toxicity, waste, and atom economy. Throughout the semester, students were asked to monitor the 12 principles from the standpoint of each of these criteria. For instance, with regard to energy usage, students were asked to compare a transformation involving a heating mantle used for 60 min with a conventional microwave oven used in the 1990s. In the way of toxicity, students were asked to research the Material Safety Data Sheet (MSDS) for some compounds prior to using them in the lab. For wastes, students were instructed to scrutinize the number of trips they take to the liquid (hazardous) organic waste container versus what is allowed "down the sink drain." While each of these is emphasized throughout the lab units at the University of Detroit Mercy, it is atom economy that receives the most attention.

Using the CHM 2300 lab [5] as a prototype, the green chemistry principle of atom economy [6] is specifically highlighted in our lab courses. This is done for two reasons. First, most students will arrive at their second year with a basic understanding of the concept of percent yield. This simple calculation allows students to quickly ascertain the effectiveness of a transformation by measuring the measured versus the theoretical yield. Second, students were straightforwardly able to extend their understanding of percent yield into atom economy calculations [7]. While other metrics for mass balance of synthetic reactions exist [8–10], simple experimental atom economy calculations allow students to immediately evaluate matters of wasted reagent mass, scale, and effects of catalysts using relatively uncomplicated algebra. With atom economy ratings, students in CHM 2300 can be asked to compare the "greenness" of one transformation versus another or suggest areas to improve the overall adherence of a transformation to the 12 principles. These calculations are simple enough that they can be templated (Figure 7.1) so that students spend less time manipulating numbers and more time relating the meaning of the ratings with regard to green chemistry.

The CHM 2300 course involves a number of complex learning outcomes in the areas of writing/editing scientific manuscripts and advanced organic synthetic techniques. However, students who fully engage the outcomes with regard to green chemistry and atom economy are more likely to overlay this important way of thinking with their existing chemical knowledge. Such students can glean the impact of their benchwork with a reasonable amount of detail in a very short period of time.

7.3 Green Chemistry Principles Affect Materials and Equipment

In examining the 12 principles of green chemistry, reform in the curriculum of our organic chemistry lab courses at the University of Detroit Mercy also affected the equipment and materials that are stocked. As previously mentioned, the lab space used for instruction in these courses contained limitations with regard to storage space, fume hood usage, and benchtop work areas. In addition, after a partial building renovation in 2011 that included a decentralization of the department stockroom, more fine chemicals and solvents would need to be stored within the teaching lab proper. Perhaps the most marked change in the way the organic lab courses are taught that followed from these factors is the overall reduction in the type and amount of solvents kept on hand. With reduced storage room, space for large (ca. 20 L) solvent cans to be stored safely had to be meted out carefully. Only keeping the amount of solvent needed for

CHM 2300 Winter 2016
ORGANIC CHEMISTRY LABORATORY II

Handout 10B
Reference – Example of Green Chemistry Calculations

Synthesis of Aspirin
Green Chemistry Calculation Sheet

Theoretical yield (TY)
Reagent table

Reagent	MM	Mass used (g)	Moles used	Moles needed	Density (g/mL)
Salicylic acid	138	1.38 g	0.01	0.01	N/A
Acetic anhydride	102	3.06 g	0.03	0.01	1.08
Phosphoric acid	98	One drop	?	Catalytic	1.685

Product table

Product	MM	Theoretical yield (mol)	Theoretical yield (g)	Actual yield	%Yield
Aspirin	180	0.01	1.80 g	1.62 g	90%

TY = (mol limiting reagent)(product to limiting reagent stoichiometric ratio)(MM of product) = 1.80 g

Percent yield (PY)
PY = (actual yield/theoretical yield) × 100 = 1.62 g/1.80 g = 90%

Percent atom economy (PAE)
Atom economy table

Reagent	MM	Utilized atoms	Mass of UA (g)	Unutilized atoms	Mass of UUA (g)
Salicylic acid	138	7C, 5H, 3O	137	H	1
Acetic anhydride	102	2C, 3H, O	43	2C, 3H, 2O	59
Phosphoric acid	98	N/A	0	P, 4O, 3H	98
Total	338	9C, 8H, 4O	180	P, 2C, 6O, 7H	158

PAE = (mass utilized atoms/total mass all reactants) × 100 = 180 g/338 g = 53%

Percent experimental atom economy table (PEAE)
Experimental atom economy table

Reagent	Moles/mass (g) used	Utilized atoms	Mass of UA (g)	Unutilized atoms	Mass of UUA (g)
Salicylic acid	0.01/1.38	7C, 5H, 3O	1.37	H	0.01
Acetic anhydride	0.03/3.06	2C, 3H, O	0.43	2C, 3H, 2O	2.63 (XS)
Phosphoric acid	?/1 drop	N/A	0	P, 4O, 3H	?
Total	0.04/4.44	9C, 8H, 4O	1.80	P, 2C, 6O, 7H	2.64

PEAE = (theoretical yield/total mass of all reactants) × 100 = (1.80 g/4.44 g) × 100 = 41%

PY × EAE
PY × EAE = (actual yield/total mass of all the reactants) × 100 = (1.62 g/4.44 g) × 100 = 36%

Overall "greenness"
Fair to poor

Figure 7.1: Example atom economy calculation sheet from CHM 2300.

the academic year can be problematic, but with proper planning, it can be realized. We also decided to reduce the total types of stored bulk solvents down to hexanes, ethyl acetate, methylene chloride, and some diethyl ether. Alcohols were pared down to methanol and ethanol only, and these solvents were purchased in different-sized containers so as to purposely exhaust the alcohol during the year.

A move was also made to using digital thermocouples with wire probes, which allowed us to eschew the use of mercury thermometers in the lab. Digital melting point apparatus, while expensive, can be another way to remove mercury thermometers from lab use. Experiments involving “grocery reagents” like caffeine [11], spinach [12], and terpenoids [13] helped reduce cost and deeply depress student safety concerns. Each semester, these items are incorporated into students’ instruction in the area of green chemistry as specific examples of reducing waste, toxicity, and hazards while increasing the safety of the learning environment. No changes with regard to equipment and materials were made without specifically choosing to enhance the learning and safety of the students.

7.4 Green Chemistry Principles Affect the Transformations Performed

The true power of organic chemistry lies in its ability to generate new structures through synthesis. Many of the green chemistry principles are tied directly to the methods of synthesis, predominantly from the perspective of energy efficiency, preventing waste, and increasing atom economy. With regard to the lab course learning outcomes already discussed, nearly all of the methods practiced and transformations performed in the University of Detroit Mercy organic lab sequence have been “greened” in several different ways. First, the 1980s and 1990s saw the arrival of microscale glassware and preparations. When these concepts first reached the academic teaching setting, they were mostly driven by cost and safety. The modern lab can take advantage of microscale work in efforts to adopt the 12 principles of green chemistry. Scale affects the waste generated, and our labs take this fact into consideration for each experiment performed. Also, over the length of the semester, total waste has been reduced significantly by teaching students about how atom economy is affected by transformation scale [7].

Second, there can be no doubt that microwave irradiation as a method of heating a reaction mixture has come into its own over the last 10–15 years, especially in the teaching laboratory. However, even a smaller microwave reactor apparatus can be expensive, and student throughput is of utmost significance at primarily undergraduate institutions, hence the advent of many reactions that reduce reaction times to hundreds of seconds (or less) by using nonfood-rated conventional microwave ovens [14, 15]. As long as these “reactors” are used in vented environments, students can compare energy usage and time of reaction between more green and less green transformations.

Third, transformations have been specifically chosen due to their solvent-free, water-based, or mechanochemical nature [16–18]. The pedagogical, educational literature has exploded with laboratory experiments that summon the most fundamental of green chemical applications: the exclusion of solvent from atom economy.

A number of transformations forgo solvent altogether or use renewable resources like water to increase atom economy and overall "greenness" for a reaction. In addition, with this motif in mind, students can learn about more atypical reaction schemes, such as those that result from added mechanical pressure, like mechanochemistry.

Lastly, a cognizance of green chemistry in modern synthesis is taught in the organic labs at the University of Detroit Mercy using click, Suzuki, and Sonogashira couplings that rely on greened catalysts [19–21]. Again, using the metrics of atom economy, students can ascertain that small amounts of compounds that undergo transformations many times are beneficial to the overall greenness of a reaction. At the University of Detroit Mercy, we have even used these cycloadditions and cross-couplings, in conjunction with traditional, less green versions, to help students compare legacy conditions to new, greener ones. Each of the items -mentioned earlier benefits the student by way of learning outcome enhancement and safety, not to mention the general cost savings of running smaller, greener experiments for many students concurrently.

7.5 Conclusions

In summary, an initiative to incorporate the 12 principles of green chemistry in the organic chemistry laboratory courses at the University of Detroit Mercy has had multiple positive effects. From helping students learn how to merge green chemistry theory with their existing chemical knowledge, to altering the overall safety profile of the lab, to cementing application of the principles through microwave, solvent free, and green catalyst experimentation, students at Detroit Mercy are well poised to be the socially responsible scientists demanded by the modern scientific workplace.

References

[1] Lab Safety Institute. labsafetyinstitute.org (accessed March 1, 2016).
[2] American Chemical Society Green Chemistry Institute. acs.org/content/acs/en/greenchemistry (accessed March 1, 2016).
[3] American Chemical Society Startegic Plan for 2016 and Beyond. strategy.acs.org (accessed March 1, 2016).
[4] Anastas, P. T.; Warner, J. C. *Green Chemistry: Theory and Practice*. Oxford University Press: New York, 1998, p. 30.
[5] Beltman, R. J.; Dierker, T. M.; Fei, A. G.; Fuchs, W. K.; Katsimpalis, A. M.; Ponkowski, M. J.; Wong, R.; Mio, M. J. An Advanced Organic Chemistry Laboratory Course Incorporating Writing/Reviewing Scientific Manuscripts and Green Chemistry Metrics, in *Environmental Chemistry: Pedagogical Methods and Practices*, Benvenuto, M. A.; Evans, K. R.; Lanigan, K. C.; Rihana, A.; Roberts-Kirchhoff, E. S. (eds.). ACS Symposium Series 1214; American Chemical Society: Washington, DC, 2015, p. 91–104.
[6] Trost, B. M. The Atom Economy – A Search for Synthetic Efficiency. *Science* 1991, *254*, 1471–1477.
[7] Green Module for Organic Chemistry at the University of Scranton. scranton.edu/faculty/cannm/green-chemistry/english/organicmodule.shtml (accessed March 1, 2016).

[8] Andraos, J.; Sayed, M. On the Use of “Green” Metrics in the Undergraduate Organic Chemistry Lecture and Lab to Assess the Mass Efficiency of Organic Reactions. *J. Chem. Educ.* 2007, *84*, 1004–1010.

[9] Ribeiro, M. G. T. C.; Machado, A. A. S. C. Holistic Metrics for Assessment of the Greenness of Chemical Reactions in the Context of Chemical Education. *J. Chem. Educ.* 2013, *90*, 432–439.

[10] Ribeiro, M. G. T. C.; Yunes, S. F.; Machado, A. A. S. C. Assessing the Greenness of Chemical Reactions in the Laboratory Using Updated Holistic Graphic Metrics based on the Globally Harmonized Systems of Classification and Labeling of Chemicals. *J. Chem. Educ.* 2014, *91*, 1901–1908.

[11] Murray, S. D.; Hansen, P. J. The Extraction of Caffeine from Tea: An Old Undergraduate Experiment Revisited. *J. Chem. Educ.* 1995, *72*, 851.

[12] Quach, H. T.; Steeper, R. L.; Griffin, G. W. An Improved Method for the Extraction and Thin-Layer Chromatography of Chlorophyll a and b from Spinach. *J. Chem. Educ.* 2004, *81*, 385–387.

[13] Kjonaas, R. A.; Mattingly, S. P. Acid-Catalyzed Isomerization of Carvone to Carvacrol. *J. Chem. Educ.* 2005, *82*, 1813–1814.

[14] Mirafzal, G. A.; Summer, J. M. Microwave Irradiation Reactions: Synthesis of Analgesic Drugs. *J. Chem. Educ.* 2000, *77*, 356–357.

[15] (a) van der Berg, A. W. C.; Hanefeld, U. 4-Dimethylaminopyridine or Acid-Catalyzed Syntheses of Esters: A Comparison. *J. Chem. Educ.* 2006, *83*, 292–293; (b) Bari, S. S.; Bose, A. K.; Chaudhary, A. G.; Manhas, M. S.; Raju, V. S.; Robb, E. W. Reactions Accelerated by Microwave Radiation in the Undergraduate Organic Laboratory. *J. Chem. Educ.* 1992, *69*, 938–939.

[16] Touchette, K. M. Reductive Amination: A Remarkable Experiment for the Organic Laboratory. *J. Chem. Educ.* 2006, *83*, 929–930.

[17] (a) Leung, S. H.; Angel, S. A. Solvent-Free Wittig Reaction: A Green Organic Chemistry Laboratory Experiment. *J. Chem. Educ.* 2004, *81*, 1492–1493; (b) Morsch, L.A.; Deak, L.; Tiburzi, D.; Schuster, H.; Meyer, B. Green Aqueous Wittig Reaction: Teaching Green Chemistry in Organic Teaching Laboratories. *J. Chem. Educ.* 2014, *91*, 611–614.

[18] Palleros, D. R. Solvent-Free Synthesis of Chalcones. *J. Chem. Educ.* 2004, *81*, 1345–1347.

[19] Hill, N. J.; Bowman, M. D.; Esselman, B. J.; Byron, S. D.; Kreitinger, J.; Leadbeater, N. E. Ligand-Free Suzuki–Miyaura Coupling Reactions Using an Inexpensive Aqueous Palladium Source: A Synthetic and Computational Exercise for the Undergraduate Organic Chemistry Laboratory. *J. Chem. Educ.* 2014, *91*, 1054–1057.

[20] (a) Lipshutz, B. H.; Chung, D. W.; Rich, D. Sonogashira Couplings of Aryl Bromides: Room Temperature, Water Only, No Copper. *Org Lett* 2008, *10*, 3793–3796; (b) Lipshutz, B. H.; Ghorai, S. Transition Metal Catalyzed Cross-Couplings Going Green: in Water at Room Temperature. *Aldrichimica Acta* 2008, *41*, 59–72; (c) Amatore, C.; Blart, E.; Genet, J. P.; Jutand, A.; Lemaire-Andoire, S.; Savignec, M. New Synthetic Applications of Water-Soluble Acetate Pd/TPPTS Catalyst Generated In Situ. Evidence for a True Pd(0) Species Intermediate. *J Org Chem* 1995, *60*, 6829–6839.

[21] Sharpless, W. D.; Wu, P.; Hansen, T. V.; Lindberg, J. G. Just Click It: Undergraduate Procedures for the Copper(I)-Catalyzed Formation of 1,2,3-Triazoles from Azides and Terminal Acetylenes. *J. Chem. Educ.* 2005, *82*, 1833–1836.

Meghna Dilip and Margaret E. Kerr

8 Greening the Curriculum: Traditional and Online Offerings for Science and Nonscience Majors

Abstract: This chapter describes the efforts of the faculty at Worcester State University (WSU) to infuse green chemistry into the undergraduate chemistry curriculum. It specifically focuses on the structure of two stand-alone classes. One is aimed at a chemistry audience and is an upper-level elective for the major. The other is an online course aimed at a nonscience audience. Both are three-credit lecture-only classes.

Keywords: green chemistry, sustainability, chemical education, online courses, undergraduate course design

8.1 Introduction

Since the publication of the 12 principles of green chemistry, education of the next generation of chemists has been seen as a cornerstone to the acceptance and development of green chemistry as a field [1]. To this end, several courses have been offered at various universities in the United States and Canada in different formats: (1) as part of an existing course; (2) as a seminar; (3) and as a stand-alone course [2–6]. An online green chemistry course is also available through Carnegie Mellon University [7]. Some classes have a focus on nonmajors [8, 9]. Several resources currently exist for greening organic chemistry labs [10–13] but fewer exist for other fields of chemistry. Despite the above-mentioned examples, it is important to document as many successful attempts at various levels so as to promote the wider adoption of green chemistry in undergraduate curricula.

Worcester State University (WSU) is a 4-year primarily undergraduate institution located in Worcester, MA. The chemistry department offers both a major and a minor, and currently has around 150 registered majors. As part of the chemistry major, students are expected to complete a minimum of 45 credits of chemistry coursework, of which 12 credits are elective credits. Green chemistry concepts were first introduced at WSU in the organic laboratory sequence in 2003. The labs performed were adapted from Doxsee and Hutchison's [15] green organic lab manual. Since that time, we have created a green and environmental chemistry concentration within the chemistry major (Figure 8.1), which requires students to take an upper-level green chemistry course, environmental chemistry, and a choice of two related electives. We offer an environmental toxicology course as well as a green/environmental laboratory course. We believe that we have strong offerings for our science majors and continue to infuse green chemistry throughout the chemistry curriculum. Additionally, to fill a perceived need for students who are not science majors, we have developed

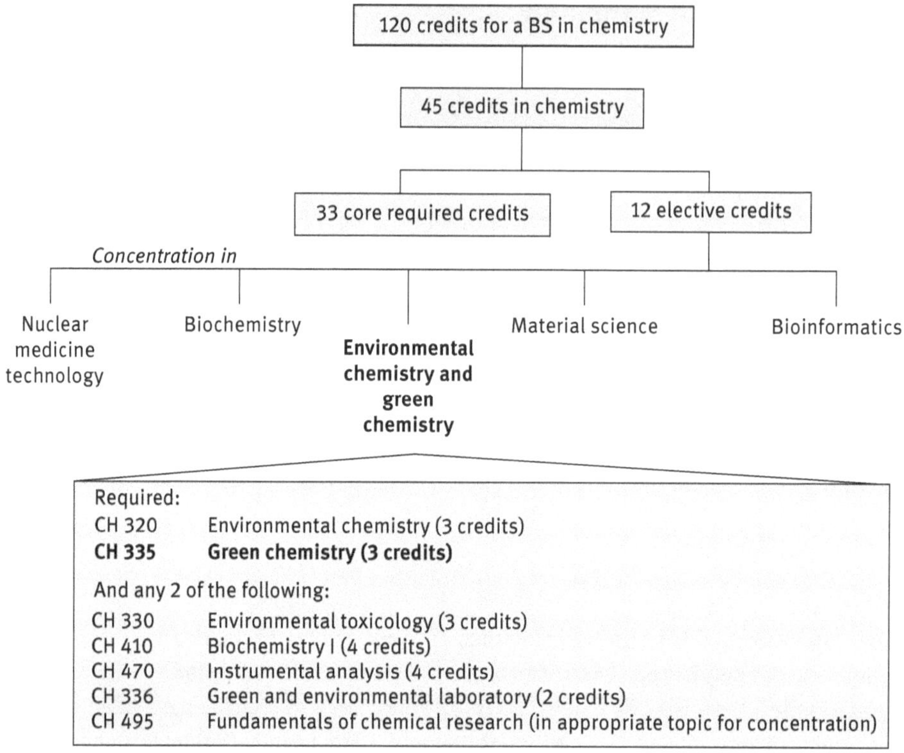

Figure 8.1: The chemistry curriculum at WSU and the "green and environmental chemistry" concentration.

an online course covering green chemistry and sustainability topics. This chapter will provide a template for our stand-alone green chemistry course for chemistry and environmental science majors and our online course on green chemistry and sustainability.

8.2 Green Chemistry – Upper-Level Chemistry Course

Green chemistry has been taught at Worcester State biannually since the spring of 2009. Both in the spring of 2009 and 2010, the class was taught with a lab. However, this did not do justice to the lab component and this class is now taught as lecture only. The class had an enrollment of approximately 25 students every time it was offered. The prerequisites for this course include a one-year sequence of general chemistry and organic chemistry as a corequisite. Students taking this class span the science spectrum and include chemistry, biology, biotechnology, and environmental science majors. The course addresses the 12 principles of green chemistry through the use of journal articles.

8.2.1 Course Outline

Introduction to green chemistry. Introduction to the 12 principles of green chemistry.
Difference between sustainability and green chemistry
Atom economy and other green metrics
Green metrics and solid state reactions/solventless reactions
Aqueous reactions
Alternative solvent media scCO_2, polymers, fluorous solvents
Green analytical chemistry – focus on chromatography
Ionic liquids – a case study
Catalysis as green chemistry and examples in biocatalysis
Case study: adipic acid synthesis
Topics in toxicology. Designing for biodegradability/Biodegradability vs. toxicology
Use of renewable materials for chemistry/renewable vs. sustainable
Alternate energy/energy considerations for reactions
Tools of green chemistry: microwaves, sonication, photochemistry

8.2.2 Course Format and Development

No textbook was used or recommended for this class. The reason for this is the author's opinion that the state of the art in green chemistry is ever-evolving. Therefore, all topics were taught from current peer-reviewed literature. Second, teaching from the peer-reviewed literature satisfies a key need for higher order thinking in the undergraduate curriculum. Green chemistry makes peer-reviewed literature accessible to students at that level. Sometimes, transcripts from the green chemistry presidential awards and Nobel prizes were used. Literature reading was emphasized and assessed throughout, including a final exam with 50 % dedicated to paper reading and questions. Students were provided with the reference to a research article in the appropriate topic. Students were usually given 1 week to find and read the article. Initially, the paper structure was discussed in class and reinforced through presentations by the instructor on the topic. As time progressed, students were given an article and asked to answer questions based on the reading before coming to class. Toward the end of the course, students were given a paper and an in-class quiz was given on their reading of the paper. This structured approach allowed the students to be gradually introduced to reading full research papers. At the end of the semester, almost all the students attested to their increase in confidence when it came to reading peer-reviewed literature. For the final, in addition to a traditional exam, students were asked to write an abstract for a paper, and answer questions based on their reading of two articles they had never seen before but based on topics covered in class. The performance on the final indicated that students did indeed feel comfortable reading and understanding literature especially with respect to the 12 principles, as indicated by them in the end-of-semester surveys.

One of the difficulties for both the student and the instructor while teaching green chemistry is the various types of chemistry spanning various fields that are presented in a single class. To reduce switching topics every class and make the paper reading more area focused, the life cycle of ionic liquids [16] and adipic acid synthesis [17] were chosen as case studies. Ionic liquids were initially hailed as green but subsequent research has shown that as a class of solvents these may or may not be green. This case study allows for a very detailed discussion of “greenness” as it applies through the life cycle of a product. Adipic acid is a raw material for the synthesis of nylon and provides an interesting industry-based case study. Using these subareas, papers were chosen for the more broad topics of catalysis, toxicity, and biodegradability. The use of case studies also enabled discussion of relative “greenness” of various synthetic strategies and comparisons. In order to make more focused comparisons particularly with regard to biodegradability and toxicity, students were introduced to databases and trained to use them effectively. Energy and economic considerations were stressed throughout. Additionally, not all reactions discussed were necessarily completely green, and students were trained to look at a process holistically. Green comparisons between various ways to carry out the same process were discussed and tested. The students had to use their judgment skills to carry out these comparisons and this was really different from the skills they are usually exposed to in other chemistry classes. Students were able to understand that there are several shades of “green,” and one process while effective in satisfying one or more principles of green chemistry may fail in other areas. They were continuously challenged to make judgments about the various options and came to gain some perspective about the economic and industrial barriers to adoption of greener techniques.

8.2.3 Modes of Assessment

The class was assessed based on regular in class quizzes (25 %) and two take-home exams (25 %). Fifty percent of the second exam was dedicated to reading two papers that the students had previously not seen and answering questions based on reading as well as writing a green context for the articles when abstract was not provided. In addition, students had to write a final paper (a literature review on any topic with a green chemistry focus) worth 25 %.

A major thrust of the course was to emphasize both written and oral communication. This is particularly important for a field like green chemistry, where spreading the word is as important as being a conscientious chemist. In general, it is known that the perception of chemists among the public is not very favorable, and concerns about pollution have contributed to this image. Better communication of green chemistry will help change this image [16]. In order to address this need, training students to communicate science and green chemistry in particular is a key goal of this course.

Several assignments (20 %) were designed with this in mind, the most novel of these being collaboration with a communications class. A professor from communications science was invited to guest lecture on the assignment at hand and particularly communicating science to a nonscience audience. Every student in the green chemistry class was paired with a student in an introductory communications class. Students in the green chemistry class were interviewed by their counterpart in the communications class. The interview included a profile of the student and questions about green chemistry and the source of the student's interest in green chemistry. At the end of the assignment, the communications major published a "biobox" of the student interviewed, and the green chemistry students wrote a reflection of the interview and their ability to communicate the concept of green chemistry to the uninformed nonscience major. Among other assignments, students were asked to write two blogs through the semester. The blogs were based on newspaper articles, other blogs, and popular media. Twitter handles were provided to students for easy discovery. Most blogs focused on energy-related issues although these were not stressed in class. This indicates a prevalence of these articles in popular media and perhaps an instinctive connection between energy and green chemistry made by the students. Students were also asked to make a presentation as a group, on a Presidential Green Chemistry Award of their choice. Here the focus was on collecting company profiles and consulting the patent literature rather than peer-reviewed publications. The use of a variety of tools to search various forms of literature in this class rather than merely focusing on a textbook allowed for students to realize the breadth of information available to them. It also allowed them to develop and hone the skills necessary to locate this information more effectively and efficiently. This instructor believes that these skills are invaluable and will lead to more skilled lifelong learners.

8.2.4 Student Survey Results

When surveyed on interest and understanding of journal articles after this class, the student responses were very positive with a large majority of students ranking their interest between 3 and 4 on a scale of 5. It was even more encouraging that most students rated their understanding of journal articles between 4 and 5. Increased student confidence in searching, retrieving, and understanding literature is a very positive outcome not only for this course but for the entire chemistry curriculum.

This increase in interest and confidence was also reflected in the final paper assignment at the end of the semester. It is the opinion of this instructor that the quality of the final papers was much better than previous iterations of this class and included the use of journal articles. On average, the papers were better researched and reflected a better understanding of the subject.

According to the student survey, interest in the specific communication assignment discussed were low with a majority of scoring their interest between 1 and 3. This is to be expected as many chemists are outside their comfort zone when it comes to communication. This is an important skill as discussed before, so this instructor does not see these ratings as negative. Some student responses are shown below:

- "I think it was more beneficial to use articles instead of a textbook and I feel as if I learned important skills to help me better understand them which is helpful for all science classes."
- "I realized that industry has economical advantages to adopting green chemistry."
- "Applications helped me understand how it affected my actual life + industry motivations."

8.3 "Paper or Plastic?": Online Approach for Nonmajors

A course titled "Paper or Plastic?" has been offered online as a nonlaboratory science elective since 2012. It has been offered a total of six terms as of this writing, with course enrollments for each section capped at 25. Due to high demand for online offerings, particularly those fulfilling science electives, each term has had more than one section available. Because many of the students taking this course to fulfill their science elective do not have a strong chemistry background, it was difficult to set the course up along the 12 principles of green chemistry as other courses have done [18]. Instead current issues involving green chemistry and sustainability such as climate change, pollution, polymers, toxicity, chemicals in common consumer products, sustainability, and so on are covered in separate blocks. Students are taught very basic chemistry principles at the beginning of the course and within each block, the chemistry behind the major topics within these issues are discussed. When possible, green chemistry and/or sustainable alternatives are presented. By the end of the course, students are expected to apply basic scientific principles to these different topics. It is expected that students will develop a basic understanding about green chemistry and greener alternatives to global environmental problems. While there are similar courses being taught at different institutions [7, 9, 19], this is the first example of a fully online course on green chemistry and sustainability targeting nonscience students.

8.3.1 Course Outline

Block 1: Green chemistry and sustainability

- What are green chemistry and sustainability and how are they different?

- How leaders in the industry are changing; how chemistry is done; and why is it important for the future?
- Consumerism/purchasing choices
- Global issues (this course is cross-listed with our Global Studies program)

Block 2: Fundamentals of chemistry
- Atomic structure
- Bonding and structure of molecules
- Organic molecules
- General reactivity of different molecules
- Unit Conversions
  - Basic discussion of metrics, scientific notation, and measurements

Blocks 3 and 4: Climate change
- Scientific description of climate change
- What are greenhouse gases: prevalence and potential
- Relating current weather phenomenon to climate change
- What are some solutions?

Block 5: Pollution
- What are some of the chemicals causing pollution and how are they formed?
- Indoor versus outdoor pollution
- Green chemistry solutions
- How does it differ from climate change?

Block 6: Chemicals in consumer products
- Identification of chemical hazards
- Examples to include bisphenol A (BPA), dioxin, polychlorinated biphenyls and others
- What are some greener alternatives?
- Why is it so difficult to make changes to existing products?

Block 7: Endocrine disruptors
- What are they and should we be worried about them?
- What are some greener alternatives to current common consumer products?

Block 8: Polymers
- What are polymers and what makes them different from each other?
- Recycling
- Which plastics are known to leach BPA and phthalates and which ones do not?
- Potential and downside of greener plastics

Block 9: Alternative fuel and energy sources
- Where do we get our fuel energy?

- Chemistry of fuel and energy
- Alternatives to our current fuel and energy sources

Block 10: Plastics in the environment
- Consumer use, recycling, and end-of-life disposal
- Environmental issues
- Plastics to fuel?

8.3.2 Course Format and Development

Since this class is fully online, a structure of instruction and assessment had to be created. Each week had a scheduled day to release new material, and students were given a week to complete assignments. All modes were untimed, with the exception of set weekly due dates. Note that there was not a text used for this course, although an online source for a general chemistry text was provided as a reference. Online education literature indicates that a variety of forms of material presentation and assessment modes are necessary for an effective course [20]. To that end, the following methods were developed:

8.3.3 Presentation of Material

Audio lectures were created using iPad apps (Explain Everything, ShowMe). The apps allow for annotation on a PowerPoint slide coupled with audio, which provides a lecture that is similar to what would be given in a face-to-face setting. Students are also given the PowerPoint files with no audio in addition to the audio lectures. A student survey given at the end of a term indicated that 73 % of students found the audio lectures helpful (n = 30). Only 3 % of those responding did not utilize them. Each block of material had one or more audio lecture associated with it. In order to maximize student utilization, each lecture was limited to roughly 15 min. If more time was necessary to explain the topic, more than one lecture was created.

Readings were provided from a variety of sources. Since many of the topics are complicated scientifically, readings had to be chosen with care in order to provide information but not overwhelm. Appropriate level readings were found in *Science*, *Nature*, *C&EN*, *the New York Times Science* section, and other media. A balance was attempted between more advanced scientific articles and ones that related topics to current events or issues. From the student survey, 90 % read all or most of the materials provided.

Online sources such as *TED Talks* and documentaries were provided if there were appropriate resources available for the topic in question.

8.3.4 Modes of Assessment

Quizzes were given as part of each block. Quizzes ranged from basic multiple-choice questions from the readings/lectures to essays that required students to apply the concepts from the block.

Discussion boards were given for those topics where it seemed appropriate for students to discuss their ideas and questions. An introductory discussion board was developed for the first block so students could get to know each other and make a connection with other students. This was not done initially and was suggested after the end of the first online semester by students. It has helped "break the ice" for students to discuss their major, why they are in the class, their knowledge of the environment and to indicate whether they have worries about taking a science course. Other discussion boards required students to have completed the readings and quiz for the block and to answer a set of posed questions. Each student was required to provide a comment of substance and to respond to two other students in order to receive full credit.

A mid-term exam was given that was included short answer and multiple-choice questions that related the first part of the course materials. There was a short essay question that required students to read an article and apply their knowledge to a set of questions. Due to the nature of an online class and information accessibility, all prior materials were left "live" so students could access them to take the exam. Questions were designed so that students had to apply their knowledge rather than hunt through the materials to find the correct answer.

A final paper was assigned that allowed students the opportunity to answer the question of the course: Paper or Plastic? Students are asked to do a life cycle assessment (LCA) to analyze whether or not paper shopping bags are preferable to plastic shopping bags. This is one LCA that has been done fairly extensively online with resources available suitable for an introductory-level course audience. They are asked to come up with an answer to this question and are not allowed to fall back on the obvious solution of just using reusable bags in place of either.

A final exam is given once students have completed the paper assignment. Since the LCA of paper versus plastic bags favors plastic bags, many students expressed considerable surprise because they have generally been under the impression that paper was favorable. The question of why different communities are banning plastic bags if they are so good inevitably comes up. In order to get them to see the issue in broader terms, they are required to watch a documentary on plastics in order to dissect what is true and what is overstated based on what they have learned [21]. They are asked to discuss how their thinking has changed based on what they have learned and how they think about the topic of plastics differently after taking this course. They are asked as consumers how they viewed the documentary now with their new knowledge. At the end, they are asked to evaluate their answer to the LCA in a broader context that would include reusable bags, different types of plastics, and so on. The exam is designed to be more of a reflection on learning and changes in thinking rather than a right versus wrong assessment.

8.3.5 Student Survey Results

From the survey, 93 % of the students had taken an online class before. The majors included students from psychology, elementary education, communications, business, and other humanities. They indicated that the level of the class was appropriate for an introductory level and also that the workload was consistent with that level. Standards for online courses require that students have equal time on task that face-to-face course offers [22, 23]. For this course, the design was such that they would spend 3 h per week on the materials with extra time for homework/assessment. Again, the survey indicated that what was provided gave a similar experience online as they would receive from a face-to-face class.

Throughout the term, quiz and discussion questions are given that allow the students to demonstrate a growth in knowledge and awareness of environmental issues. A formal assessment comparing student knowledge at the end of the term versus the beginning has not yet been done, but student comments throughout the term indicate an overall gain in knowledge. Samples of student responses are as follows:

- The information I have learned in this class has opened my eyes to making better choices.
- I am overwhelmed with the complexity but feel grateful that I have a starting base knowledge about these things.
- I feel more informed than I had been previously, and I will continue to read and research healthier options so that I can live a greener life.

8.4 Conclusion

New generations of chemists must definitely be trained in green chemistry. An overwhelming number of students surveyed in both formats of the course felt the course made chemistry more relevant and raised awareness of environmental issues. The applied nature of the subject and the real-world case studies do impact a student's perception of chemistry in a positive manner.

The use of current literature is particularly important in a field such as green chemistry, where new research is constantly appearing and spans various areas of chemistry where the instructor may not be well versed. While textbooks are focused more on organic chemistry, peer-reviewed literature spans a larger variety of topics and allows for the instructor to model the class based on what may be currently capturing public interest as well as student interest. This approach was used in the green chemistry class for majors and allowed for the class to be more student centered, and leads to more engaged discussion from the students in general. The students themselves did not find the lack of textbook an impediment when surveyed.

Green chemistry is more accessible than traditional chemistry to nonmajors. Green chemistry also allows for communication between diverse departments, and a wider dissemination of chemistry within the college campus.

As instructors, there are some barriers to overcome when teaching an ever-evolving subject. Several of us have not been formally trained in green chemistry, and it is hard to define the boundaries of green chemistry. Therefore, more templates are needed for wider adoption. We hope this chapter inspires a few to try their hand at teaching their version of green chemistry that fits the needs of their student body and faculty member.

References

[1] Young, J. L.; Peoples, R. ConfChem Conference on Educating the Next Generation: Green and Sustainable Chemistry—Education Resources from the ACS Green Chemistry Institute. *J Chem Educ* 2013, *90*, 513–514.

[2] Marteel-Parrish, A. E. Teaching Green and Sustainable Chemistry: A Revised One-Semester Course Based on Inspirations and Challenges. *J Chem Educ* 2014, *91* (7), 1084–1086.

[3] Marteel-Parrish, A. E. Toward the Greening of Our Minds: A New Special Topics Course. *J Chem Educ* 2007, *84*, 245–247.

[4] Andraos, J.; Dicks, A. P. Green Chemistry Teaching in Higher Education: A Review of Effective Practices. *Chem Educ Res Pract* 2012, *13*, 69–79.

[5] Cann, M. C.; Dickneider, T. D. Infusing the Chemistry Curriculum with Green Chemistry Using Real-World Examples, Web Modules, and Atom Economy in Organic Chemistry Courses. *J Chem Educ* 2004, *81*, 977–980.

[6] Machanyatage, R. Designing and Incorporating Green Chemistry Courses at a Liberal Arts College To Increase Students' Awareness and Interdisciplinary Collaborative Work. *J Chem Educ* 2013, *90*, 1167–1171.

[7] Learning Green – The Institute of Green Science. http://igs.chem.cmu.edu/ (accessed July, 2014).

[8] Gross E. M. Green Chemistry and Sustainability: An Undergraduate Course for Science and Nonscience Majors. *J Chem Educ* 2013, *90*, 429–431.

[9] Prescott, S. Green Goggles: Designing and Teaching a General Chemistry Course to Nonmajors Using a Green Chemistry Approach. *J Chem Educ* 2013, *90*, 423–428.

[10] Dicks, A.P. (ed.). *Green Organic Chemistry in Lecture and Laboratory.* CRC Press: Boca Raton, FL, 2011.

[11] Morsch, L. A.; Deak, L.; Tiburzi, D.; Schuster, H.; Meyer, B. Green Aqueous Wittig Reaction: Teaching Green Chemistry in Organic Teaching Laboratories. *J Chem Educ* 2014, *91* (4), 611–614.

[12] Burtch, E. A.; Jones-Wilson, T. M. A Green Starting Material for Electrophilic Aromatic Substitution for the Undergraduate Organic Laboratory. *J Chem Educ* 2005, *82* (4), 616–617.

[13] Goodwin, T. E. An Asymptotic Approach to the Development of a Green Organic Chemistry Laboratory. *J Chem Educ* 2004, *81* (8), 1187–1190.
[14] Roesky, H. W.; Dietmar, K. (ed.). *Experiments in Green and Sustainable Chemistry*. Wiley-VCH: Weinheim, 2003.
[15] Doxsee, K.; Hutchison, J. *Green Organic Chemistry: Strategies, Tools and Laboratory Experiments*, 1st ed., Cengage Learning: Boston, MA, 2003.
[16] Dilip, M. Cradle to grave: How green are ionic liquids? *Nanomater Energy* 2012, *1*, 193–206.
[17] The Changing Image of Chemistry, Chemistry World. December 2004, http://www.rsc.org/chemistryworld/Issues/2004/December/changingimage.asp (accessed July, 2014).
[18] Tamburini, F.; Kelly, T.; Weerapana, E.; Byers, J. A. Paper to Plastics: An Interdisciplinary Summer Outreach Project in Sustainability. *J Chem Educ* 2014, *91*, 1574–1579.
[19] Rosen, L. http://fod.msu.edu/oir/evaluating-online-courses and references therein (accessed November 2, 2015).
[20] Neue Sentimental Film, Brandstorm Entertainment, and Cine Cartoon Filmproduktion GmbH present; a Werner Boote film; produced by Thomas Bogner, Daniel Zuta; written and directed by Werner Boote. *Plastic Planet*. First Run Features: New York, NY, 2011.
[21] Dubins, B. H.; Graham, M. B. 2009. Training instructors to teach online: Research on competencies/best practices, http://www.uwex.edu/disted/conference/Resource_library/proceedings/09_20433.pdf (accessed November 2, 2015).
[22] Online and hybrid course development at Bowie State University. 2011, http://www.bowiestate.edu/files/resources/onlinehybrid_manual_2014_v8.pdf (accessed November 2, 2015).
[23] WASC Senior College and University Commission. Best practice strategies to promote academic integrity in online education, Version 2.0, June 2009, http://www.wascsenior.org/files/Best_Practices_for_Academic_Integrity_in_Online_Education.pdf (accessed November 2, 2015).

Index

Also of interest

Volume 2
Green Chemical Processes.
Benvenuto (Ed.), 2017
ISBN 978-3-11-044487-2, e-ISBN 978-3-11-044592-3
ISSN 2366-2115

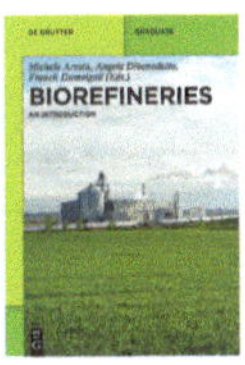

Biorefineries.
An Introduction
Aresta, Dibenedetto, Dumeignil (Eds.), 2015
ISBN 978-3-11-033153-0, e-ISBN 978-3-11-033158-5

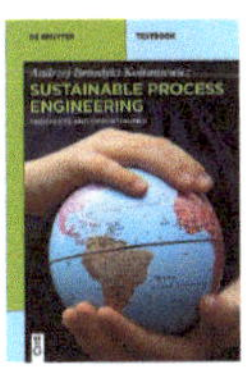

Sustainable Process Engineering.
Prospects and Opportunities
Koltuniewicz, 2014
ISBN 978-3-11-030875-4, e-ISBN 978-3-11-030876-1

Environmental Nanoscience.
Implication of Anthropogenic Nanomaterials
Obare (Ed.), 2018
ISBN 978-3-11-034234-5, e-ISBN 978-3-11-034235-2

Sustainable Chemical Production Processes.
Marin, van Geem, 2018
ISBN 978-3-11-026975-8, e-ISBN 978-3-11-026992-5

GREEN – Alternative Energy Resources
Volume 1
Pyrolysis of Biomass.
Wang, Luo, 2016
ISBN 978-3-11-037457-5, e-ISBN 978-3-11-036966-3
ISSN 2509-7237

Kolibri 360 GmbH
Ferdinand-Jühlke-Straße 7
99095 Erfurt, Deutschland
produktsicherheit@kolibri360.de